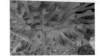

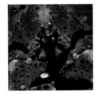

THE SHAPE OF LIFE

THE SHAPE OF LIFE

Nancy Burnett and Brad Matsen

Monterey Bay Aquarium Press
in association with Sea Studios Foundation
Monterey, California

DEDICATIONS

To Donald P. Abbott and Charles H. Baxter

Nancy Burnett

To Laara Matsen and Jonas Bendiksen

Brad Matsen

The mission of the Monterey Bay Aquarium is to inspire conservation of the world's oceans.

Sea Studios Foundation is dedicated to communicating a sense of wonder, appreciation and respect for the world around us through the lens of science.

Published in the United States by Monterey Bay Aquarium Foundation in association with Sea Studios Foundation, 886 Cannery Row, Monterey, California, 93940-1085. www.montereybayaquarium.org. The Monterey Bay Aquarium and kelp logo are registered trademarks of the aquarium.

Library of Congress Cataloging-in-Publication Data

Burnett, Nancy.
    The shape of life / by Nancy Burnett and Brad Matsen.
        p. cm.
Includes bibliographical references (p.   ).
    ISBN 1-878244-39-6
  1. Morphology (Animals).   I. Matsen, Bradford.   II. Title.
    QL799 .B87 2002
    571.3'1--dc21

                                                                        2001008465

Printed on recycled paper and bound in Hong Kong by Global Interprint.

9   8   7   6   5   4   3   2   1

Abraham Trembley text from *Hydra and the Birth of Experimental Biology, 1744: Abraham Trembley's Mémoires Concerning the Natural History of a Type of Freshwater Polyp with Arms Shaped Like Horns* reprinted by permission of the publisher, The Boxwood Press, Pacific Grove, California.

Designer: Beth Hansen-Winter
Managing Editor: Michelle McKenzie
Project Editor: Nora L. Deans
Photo Editor: Nancy Burnett
Photo Research: Stephen Ulrich

Cover image: 4:2:2 Interactive Media

ACKNOWLEDGMENTS

This book and *The Shape of Life* television series would never have been completed without the inspiration and work of many people. I owe an immense debt of gratitude to Mark Shelley and Chuck Baxter for sharing my enthusiasm and vision for *The Shape of Life* project and for building Sea Studios into a home for this kind of work. Thanks to David Elisco, the series producer, who crafted eight hours of compelling television stories, to Chuck Baxter and Tierney Thys who kept the science honest and accurate in both the series and this book, and to the producers, staff and crews, and the many scientists who contributed so much to this project.

Many friends and teachers have inspired my love for biology and the natural world, but without Don Abbott, Chuck Baxter and Ken Norris, the joy and work of science would not have flourished in my life.

Thanks to our editor, Nora Deans, our designer, Beth Hansen-Winter, our publisher, Michelle McKenzie and the rest of the staff at the Monterey Bay Aquarium Press. My children, Jason, Sierra, and Christopher tolerated their busy mother during the five years of work on this project, and Steve Webster, Susie Harris, Lisa Utall, Julie Packard, and my other good friends provided me with warmth and support. And thanks, especially, to Brad Matsen for sharing my enthusiasm for the animals and their stories and bringing his words to this book.

Nancy Burnett

Many thanks to my generous friends and colleagues without whom life and books would be impossible. At home, Kay Wilson, Kurt Esveldt, Laara Matsen, and Jonas Bendiksen took care of me. At Sea Studios, Mark Shelley provided me the great opportunity to work on *The Shape of Life* project, Chuck Baxter and Tierney Thys taught me biology, and many other people kept the whole show on the road. At the Monterey Bay Aquarium Press, Nora Deans is an editing angel, Beth Hansen-Winter designed a beautiful book, and Michelle McKenzie and her staff took care of the many details of publishing. I am especially grateful to my co-author, Nancy Burnett, who brought her immense curiosity and attention to detail to this book, along with a wonderful collaborative instinct that made the work joyful.

Brad Matsen

CONTENTS

Preface

THE SHAPE OF LIFE

— ix —

Chapter 1

ANIMAL EVE: SPONGES

— 1 —

Chapter 2

LIFE ON THE MOVE: CNIDARIANS

— 19 —

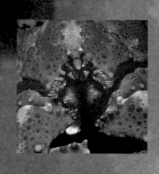

Chapter 5

THE CONQUERORS: ARTHROPODS

— 63 —

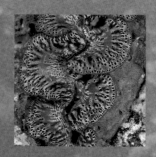

Chapter 6

THE SURVIVAL GAME: MOLLUSCS

— 77 —

Chapter 3

**THE FIRST HUNTER:
FLATWORMS**

—35—

Chapter 4

**AN EXPLOSION OF
LIFE: ANNELIDS**

—49—

Chapter 7

**AN ULTIMATE
ANIMAL?:
ECHINODERMS**

—91—

Chapter 8

**BONES, BRAWN AND
BRAINS:
CHORDATES**

—101—

Bibliography and Further Reading   —115—

Phylogenetic Tree of Life   —117—

Index   —119—

PREFACE

PREFACE

The Shape of Life

The next time you're sitting around with a bunch of people and there's a lull in the conversation, ask them to name their favorite animals. My bet is that not a single person will answer with any of my favorites, a polyclad flatworm or a purple encrusting sponge or the market squid, *Loligo opalescens*. Chances are, the lists you get will include only members of the small (though certainly not insignificant) group that includes the lions, tigers, bears, and us. I have nothing against that group, of course, but I just happen to be partial to the 95 percent of the animal kingdom that comes without a backbone.

My love affair with invertebrates started with a limpet when I was a kid on our family vacation in Carmel, California. I was exploring a tide pool and pried off what I thought was a chip of rock to discover that it came attached to something soft underneath that was a beautiful orange color with a headlike thing on one end. I was amazed. It wasn't until years later when I took a biology course that I found out that limpets, like the one I'd picked up as a child, actually have very complex lives and some interesting relatives. At regular intervals, I learned, a limpet ventures out to scrape food off the rocks with a special tongue called a radula. Amazing. But what was most incredible to me was that it returns to exactly the same spot on the rock to rest until its next feeding foray. Years later, I had the opportunity to focus a camera on a rock full of limpets. Click, click, click. When that sequence of stills taken over 24 hours was strung together and I actually watched the limpet waltz around on the rocks, I was absolutely floored. And hooked forever on creatures that most people don't even consider to be animals.

Fortunately, my life brought me friends and colleagues who share my fascination, particularly a couple of biology fanatics named Chuck Baxter and Nancy Burnett. Our paths intersected at Stanford 30 years ago, and we have been working and playing together ever since, most of the time under the supervision of animals. We launched Sea Studios, participated in the creation of the Monterey Bay Aquarium, and during one joyful weekend in 1995, decided that we wanted to celebrate the animal kingdom in a television series. Brad Matsen joined us soon after, bringing his own passions for words and scientific inquiry to the project. We knew the animals had great stories to tell, we knew they enriched our lives, and we wanted to share our good fortune with as many people as possible.

The Shape of Life television series and this book are the results of that decision and six years of dedication by hundreds of people. But the animals themselves are the stars, and they also organized our work for us. Every animal that ever lived fits into one of 35 basic body plans that evolved more than half-a-billion years ago and endure to this day. More than 95 percent of all animals fit into just eight of those *bauplans*, so the series has eight episodes, and this book has eight chapters, each of which focuses on one of them. In telling their stories, the animals make it clear that none of them is 'better' than any other; that evolution is about adequacy rather than perfection; and that diversity is the key to the survival of living things.

What is most remarkable about the journey you are about to make through the animal kingdom is that so much of it will take you through unknown territory, but therein lies the potential for the same fascination and beauty I found when I pried a limpet off a rock in a Carmel tide pool. Welcome to *The Shape of Life*.

Mark Shelley
Co-Executive Producer,
Sea Studios and National Geographic Television

SPONGES

CHAPTER 1

ORIGINS

Animal Eve

I'm obsessed with learning where we came from. And what I mean by that is not who your mother and father are, or your grandparents, but rather where do the organisms come from that gave rise to complex animals like us?

Mitch Sogin, Biologist

Imagine a family reunion, maybe held in July on Tampa Bay where your grandparents have lived since you were a kid and where you spent two weeks every summer in their skiff, shirtless and fishing for sea trout, burning yourself the color of persimmons. Twenty-five years later, you arrive in a rental car from the airport with your eleven-year-old twin girls who barely tolerate the two-hour drive. They're uncertain about meeting a hundred new people, some of whom, you told them when you planned the trip, will look like them. You park the car in a sawgrass field behind the house, and follow a path through overhanging cypress and drapes of kudzu to a sandy beach where the splashes of the adults' bright shirts and Bermuda shorts mingle with the blurs of bathing-suit greens, yellows and reds of the children in constant motion.

The shapes and faces of your kin resolve slowly as you approach them standing around the picnic tables, the volleyball court and the barbecue pit. There is your grandfather, a widower, who carries so much of you in his features you know for sure what you will look like in 50 years. And there are your father and mother, aunts, uncles, their children, their children's children, some tall, some short, blondes, brunettes, even a couple of redheads, but most of them blue-eyed. As your attention shifts to handshakes and hugging, your children wade into the crowd and walk directly over to another pair of twins, two boys, with long, high-cheeked faces and blocky shoulders who look so much like your own as to be mirrors. Your daughters aren't really aware of the familiar physical cues they picked up in their cousins, but their long ears, cornflower-blue eyes, high foreheads, and sharp noses are somehow familiar, comforting, and safe.

The powerful connections that we call 'blood' obviously tie us to our kin through countless generations of love, hate, joy, sorrow, mystery and all things human. That sense of the familiar has been embroidered into our subconscious in just a few hundred years, and we can easily imagine what our ancestors looked like over that period of comprehensible time. But all time is not comprehensible. Six thousand years ago, we began recording history. Language itself has been around for only 50,000 years. Our species, *Homo sapiens*, appeared 500,000 years ago. Now try to imagine a million years, a hundred million. It's impossible, really, but only deep time contains the explanations of our ancestral connections that go beyond our familiar facial features. Grasping the reality of years

1

in the millions requires us to resort to the uniquely human power of abstraction to get past the scale of our life-spans, but only then can we begin to fathom our relationship with all other animals through time.

We know intuitively, without comprehending the true depth of time, that some thread of life ties us to our non-human ancestors. But how far back? The family album kept so carefully by your great-aunt Mae (the Irish slipped into the clan around the turn of the last century) and displayed on a card table at the reunion, contains photographs from as far back as great-great-great grandfather and grandmother, Karl-Olav and Inge, but that is as far as you can get before personal history fades into myth and legend. What then? As members of the primate family, we can imagine our ancestors back another 65 million years, as mammals 250 million, as vertebrates over a half-billion, to creatures whose hint of a backbone is enough to connect them to all of us. If we can trace our ancestors back through time for hundreds of millions of years, does the ancestry of other animals—snails, insects, worms, starfish—extend back that far?

And even more importantly, is it possible that all animals are somehow related way, way back in time?

We do share basic traits with all members of the animal kingdom: We are all made of many cells. We are all the product of the fertilization of a large egg by a smaller sperm, and from this single cell all animals transform themselves in a highly organized way into an adult body. Most animals have heads, mouths and legs or some other way of getting around organized in a body made of many cells doing specialized work but interacting to produce a functioning whole. Do these traits connect all animals in the way a family's high cheek bones and blue eyes show a common ancestry?

If animals can be related through a common ancestry, there must have been a first animal, the pioneer who began putting together the animal way of life, who first carried those traits we now define as "animalness," an incredible first being that started the process of populating Earth with the wondrous animal kingdom we see today, including ourselves.

> *The world becomes full of organisms that have what it takes to become ancestors. That, in a sentence, is Darwinism.*
>
> Richard Dawkins,
> *River Out of Eden: A Darwinian View of Life*

We have been on a quest to trace our ancestry and to find the origin of animal life since the concepts of past, present, and future flashed into the brains of the earliest human beings, creating imagination, curiosity, and wonder. Through most of our history, we invented myths and legends to explain where we came from, and most people believed that all the animals that are alive are all the types of animals there have ever been. Until we found the tools for comprehension just a few hundred years ago, the origin of animals, including us, was the stuff of magical gardens, hands reaching down from the sky to scatter islands and creatures, and natural cataclysms such as floods and earthquakes from which emerged beings like us. The idea that new species appear and old species die off in a process we call evolution was utterly beyond our imagination until the systematic inquiry of science triggered a revolution in the way we understand our own existence.

The chain of discoveries that eventually led us to draw some accurate conclusions about our origins and our relationships to other animals began when humans applied our talent for classifying

THE POROUS NATURE OF SPONGES IS EVIDENT IN TWO SPECIES OF VASE SPONGES (OPPOSITE PAGE). □ A GLOBULAR SPONGE (THIS PAGE, TOP) AND ENCRUSTING SPONGES (MIDDLE AND BOTTOM).

and comparing living things to trying to understand our ancestry. We celebrate ourselves as "The Discoverers," but we are also "The Classifiers" and this has allowed us to build some understanding of the meaningful links between the animals, plants, rocks, and everything else with which we share the Earth.

When he wasn't working on the other foundations of Western civilization, Aristotle poked around in tide pools and mused about whether the colored shapes and blossoms of the seafloor were 'sensate' or 'insensate,' 'animate' (which means "having a spirit") or 'inanimate'. He constructed elaborate charts describing the relationships between animals in terms of their internal and external similarities, in essence classifying them. When the old Greek genius couldn't figure out whether something was animate or inanimate, such as one of the colorful and puzzling clumps of life we now call sponges, he just said they were intermediates, or "in between." Until the eighteenth century, classification was pretty much a matter of just grouping things that looked, sounded, felt, or tasted alike, (or were somewhere in between). There was no one accepted system for organizing things like species or families of animals, and no one was thinking about where animals come from, because we presumed that all life was generated spontaneously or created by some kind of super-being.

CHARLES DARWIN

In the mid-1700s, a Swedish botanist named Carolus Linnaeus set up a standard method for naming organisms that remains in use today. When he was alive, about 10,000 animals had already been described, and his was an orderly filing system with species grouped within successively larger groups in a hierarchy based on common characteristics. Life is thus organized into groups defined by very specific characteristics shared by all members of the group. For instance, you are of the kingdom Animalia; phylum Chordata; subphylum Vertebrata; class Mammalia; order Primates; family Hominidae; genus *Homo*; and species *sapiens*. When we talk about animals, we usually just refer to them by their genus and species, as in *Homo sapiens*.

Though Linnaeus didn't know it at the time, his system would remain basically correct to this day because his procedure classified animals according to common features that reflect true evolutionary relationships. The criteria for placing different organisms into their place in the system have changed, but the system itself remains the same. Linnaeus was not looking for origins though, not looking for an Animal Eve because he was firmly rooted in his times and believed that all animals were created by God. His work, he thought, was to discover God's orderly design. God, Linnaeus thought, did the designing; he did the filing. Linnaeus did, however, pave the way for an enormous breakthrough in our understanding of the relationships between living things because he provided a communications channel not only among classifiers but among generations of classifiers. The inquiry and analysis of the past, therefore, could inform future discovery.

> *The time will come I believe, though I shall not live to see it, when we shall have very fairly true genealogical trees of each great kingdom of nature.*
>
> Charles Darwin, in correspondence

Then, just 150 years ago, Charles Darwin upended the way we think about the presence of life on Earth when he asked one of the greatest questions ever asked of nature: Where do species come from? He concluded correctly that the shifting demands of their environments select from the variability inherent in a population of animals to create new species. This process of natural selection could easily produce the diversity of both living and extinct species. Later, we would learn that the occasional accident plays a powerful role in evolution, too.

Darwin's celebrated voyage aboard the HMS *Beagle* was only part of a life of inquiry that led him to this answer. As with most scientific breakthroughs, Darwin's theory of evolution was an elegant synthesis of conclusions reached by early investigators and those based on his own inquiry from several perspectives. Linnaeus' classification system was of utmost importance to Darwin when he observed the incredible diversity of animal life in similar habitats around the globe. Aboard the *Beagle*, he wondered why forests in different places have different animals filling the same niches. At home in England, he bred domesticated animals and knew, among other things, that physical traits are passed on from generation to generation and can be selected by the breeder. Darwin was also a geologist and knew that the rocks and sediments he studied could have formed only over enormous amounts of time, much more time than the religious myths of his day could provide.

Darwin was a paleontologist, too, and he was able to make the connection between the fossils he collected and the living animals that are related to those ancient beasts. Although there were explanations for the existence of fossils, other investigators in the early centuries of systematic inquiry into the origins of life never factored the true enormity of time into their conclusions. They had absolutely no concept of millions and millions of years and no idea that time of such magnitude played any role at all in the creation or existence of animals in the present.

People had been finding fossils for centuries, even polishing them or making them into tools. Workmen digging the sewers of Paris found tusks and bones of giant elephantlike creatures nobody had ever seen alive. Miners in Holland found the skulls and fragments of giant reptilelike animals with paddles instead of feet, obviously swimmers where no ocean then existed. Some people thought they must be the remains of ancient animals, some of which were now extinct, but they didn't know how they were related to the beasts of land, sea and air in the present. There was a period of time they called the past, but the dimensions of the past were unknown.

Finally, when Darwin went exploring off the Pacific coast of South America, he saw flora and fauna that had adapted, literally changed form, to survive in different environments. In the Galapagos

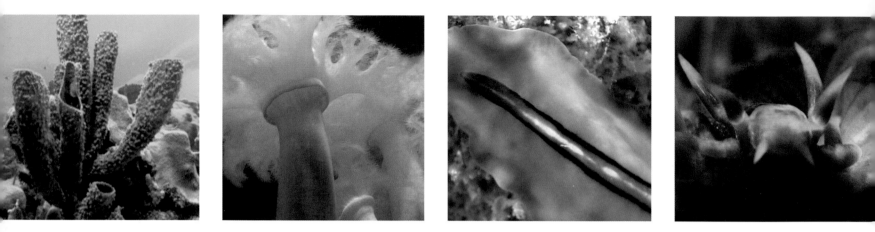

Islands, he encountered animals that were obviously similar to those living on the continent. He saw, too, that they were also different from the mainland and even different between the islands. These observations would help him put together his celebrated theory years after he returned to England. He wrote, ". . . the struggle for existence bears on natural selection." Animals that are able to secure food, avoid predators, and reproduce are those best suited for survival—the fittest. The natural processes of survival have selected them to survive and become the ancestors of subsequent generations.

So Darwin proposed a process in which one species can change through evolutionary time into another. Different species are related in just the same way we are more obviously related to our cousins, aunts, uncles, and parents. And he also knew there was an enormous amount of time available in the history of Planet Earth for evolution to take place. Though Charles Darwin, like Linnaeus, was a product of his time and desperately wanted God to have had a hand in the creation of animals, the evidence for the true origin of species was too overwhelming.

Thus, from the war of nature, from famine and death, the most exalted object which we are capable of conceiving, namely the production of higher animals, directly follows.

Charles Darwin,
The Origin of Species

Once deep time and the processes of natural selection and evolution enter into our search for our ancient ancestors, the family album gets really interesting. And so does the question of our origins. From the interlocking discoveries of the many scientific disciplines concerned with evolution, including embryology, paleontology, and genetics, we can now construct branching diagrams, called cladograms, that trace animals through time to common ancestors through physical and genetic characteristics shared by successive generations of species. Our own species, *Homo sapiens*, for instance, becomes distinct from its ancestral species, *Homo erectus*, at one of those branching points. By grouping

common traits of hominid fossils, we can move back through time from ancestor to ancestor, and from ancestral group to ancestral group.

If we go far enough back in time to a sudden, great radiation of life beginning 540 million years ago, we see the branching of animals that would lead to 35 basic body plans called phyla (the word is from the Greek, meaning simply "class.") Before that explosion of animals, we have evidence that three of those groups existed, the ancestors of spongelike, jellylike and wormlike animals. All animals today still fall into one of those 35 groups. They adapted the architecture of primitive body plans as conditions changed through time to create enormous diversity, but most animals belong to only eight of these groups.

One of the oldest members of our own phylum, the chordates, is a wormlike ocean creature named *Pikaia gracilens*, a tiny longshot that is on the first page of the family album for animals with backbones. (You can actually see a fossil of *Pikaia* on a piece of shale taken from high in the Canadian Rockies and now in a special glass case at the Smithsonian Institution.) We also have maps into the past for the other seven major body plans—sponges, cnidarians, flatworms, annelids, arthropods, molluscs, and echinoderms.

The next big question in our search for the earliest animal on Earth, of course, is how these eight branches of life are related and who is at the bottom? We know that species can beget species, so at some time in the distant past, a single phylum or body plan might have blazed the trail for all of the others. Was there a single animal from which the others might have arisen? Or did animals evolve more than once?

In the last decade of the twentieth century, with evidence from fossils and the chemistry and anatomy of living animals, some scientists searching for our origins began to suspect that the first animal on Earth, the Animal Eve, was a sponge. There was a lot of discussion, too, about whether or

THE EIGHT PHYLA INCLUDE: SPONGES, CNIDARIANS, FLATWORMS, ANNELIDS, ARTHROPODS, MOLLUSCS, ECHINODERMS, AND CHORDATES.

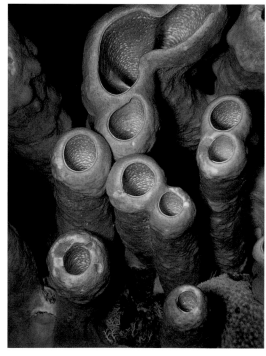

not sponges were in the direct ancestral line of modern animals. We know that sponges are composed of many cells, like all animals, and are clearly more complex creatures than the single-celled, non-animal organisms from which they must have evolved. Single cells ruled the world for billions of years. Within the group called protists would reside the ancestors of animal cells. But these cells do not clump together and cooperate like animal cells. On the other hand, sponges are made of many cells, but they, in many ways, still carry out life processes like single cells. Sponges are clearly on the edge of animalness. A sponge doesn't look anything like a horse, or a worm, or a sea star, or anything but another sponge, but it is an animal and a likely candidate to be our oldest ancestor.

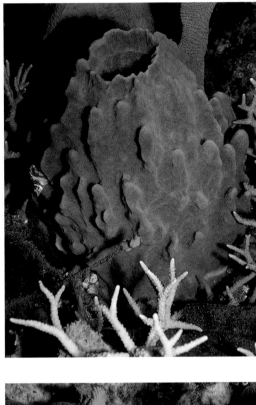

> *Nature proceeds little by little from things lifeless to animal life in such a way that it is impossible to determine the exact line of demarcation, nor on which side thereof an intermediate form should lie. So, in the sea, there are certain objects concerning which one would be at a loss to determine whether they be animal or vegetable.*
>
> Aristotle, wondering about sponges

For most of human history, though, we weren't even sure that sponges were animals. Two thousand years ago, they were listed among Aristotle's intermediates, somewhere between plants and animals. His confusion is understandable to anyone who has ever seen, but not looked too closely at, a sponge, which has no head, no brain, no bones, no mouth, and no internal organs. Sponges come in an astonishing variety of shapes that to us look like cups, fans, tubes and colorful, crusty smears on rocks and coral. They range in size from less than a half-inch wide to more than three feet tall, like the great barrel-like glass sponge that lives in Antarctic waters. All sponges are aquatic, tied permanently to the water by their lifestyle and body plan. Of the 10,000 species alive today, only 150 live in fresh water, the rest in the ocean.

Sponges officially became part of the animal kingdom in 1825 when Robert Edmund Grant, a Scottish biologist, finally convinced his peers that

A VARIETY OF SPONGE SHAPES AND COLORS INCLUDE A SPONGE GROWING IN CORAL (OPPOSITE PAGE, TOP RIGHT). CANALS RADIATE FROM THE EXCURRENT OPENING IN AN ENCRUSTING SPONGE (THIS PAGE, BOTTOM).

the creatures he studied in the tide pools and tanks were neither plants, Aristotle's intermediates, nor blobs of goo. He had long believed that sponges were animals, but finally sold everybody else when he poured colored water into a bowl of sponges to show the particles flowing into the sponge through tiny pores and then "vomiting forth from a circular cavity, an impetuous torrent." These beings moved, they pumped water through holes in their bodies; they must be animals. Grant named them phylum Porifera, which means "pore bearers." (The controversy continued, however, into the 1980s when some biology texts still did not accept sponges as true animals.)

When I encounter a sponge I'm just in awe. They're just so different you cannot stop yourself from asking: "What are you? What do you do? What do you eat?"

Christina Diaz, Biologist

The anatomy of a sponge is not as likely to inspire awe as that of a horse, a lion, a giant squid, or a marlin, which feature complex structures like tissue and organs that are very clear in their statements about form and function. Sponges, though, are a modest collection of about six different types of specialized cells that handle reproduction, food gathering, eating and digestion. (By comparison, we complex humans require over 250 different types of cells.) Their bodies don't invoke images of power, strength or classical beauty until you take a much closer look, beginning with the stunning cells of which they are built.

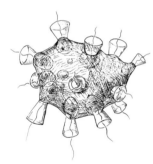

Specialized animal cells are like the individual notes of an orchestra tuning up for the Brahms "Violin Concerto," each note whole and independent but capable, with all the other notes, of becoming a complex melody. Not all living things are symphonic in this way.

The kingdom Protista is one of the great groups into which we have organized life in our attempt to understand the business of living on Earth. Each member is composed of only a single cell, although some live as colonies with perhaps two different types of cells and others appear multicellular, but they still carry out life as single cells.

All protists are just single biological notes, capable of producing only the same single note in successive generations, crawling, oozing, swimming in a vast cacophony of simple creatures. If protists as single cells came before animals, could there be some connection or transition between them and the first true animal? One of the protists, called *Proterospongia*, in fact has cells that look amazingly like one of the kinds of specialized cells present in all sponges, vigorous characters called choano-flagellates that feature collars surrounding long, whiplike appendages that are kept in constant motion. From the kingdom of protists, a choanoflagellate gives us a hint about one of the steps single cells took that moved them closer to the birth of the animal kingdom.

Animal cells, on the other hand, are capable of specializing, communicating and reorganizing

VARIOUS SINGLE-CELLED PROTISTS (TOP). THE COLONIAL CHOANOFLAGELLATE *PROTEROSPONGIA* IS A MODEL FOR THE POSSIBLE SPONGE ANCESTOR (MIDDLE). SPONGE CHOANOCYTE CELL (BOTTOM).

themselves as the animal grows. Life begins when a single cell, a fertilized egg, divides into two, then four, eight, sixteen, and on and on. If these were protists, every cell produced in this chain of creation would remain the same. Instead, animal cells differentiate according to the genetic code stored in their nuclei, and then produce succeeding generations of liver cells, or brain cells, or, in the case of a sponge, pumping cells, or whatever kind of cells define the particular animal.

Protists may be simple, but they are survivors, and one among them may have launched the world into a whole new way of organizing life. Somewhere along the line, maybe seven or eight hundred million years ago, a bunch of protist cells—probably choanoflagellates—living together as a colony began communicating so that some cells did only the reproducing and others did only the food capturing. Eventually, this way of working together would be so well established in the descendents of these adventurous ancestors that they evolved to become an entirely new life form: an animal. The single notes of the biological orchestra then began to compose the first bars of complex music.

The ensemble of specialized cells that form a sponge can pick out a tune like "Twinkle, Twinkle, Little Star," much simpler than Brahms's "Violin Concerto," but music nonetheless. The walls of a sponge's body are only three layers thick, two of which adjoin either the water around it, or one of the many canals and chambers that weave through its interior, literally bringing the ocean inside the body of the sponge. Between the layers of cells that create the boundary between the sponge and the water, inside and out, is a jellylike layer called the mesohyl, which just means "middle fluids." Several kinds of cells wander in this middle fluid, including amebocytes (crawling cells) that produce structures to support the sponge. Amebocytes also capture food particles and pass them along to the other cells. In fact, some amebocytes are capable of performing just about all the jobs required to keep a sponge alive, because they have retained the ability to transform into all the other different cell types.

Each type of cell has a specific job to do for the sponge. In a few types of sponges, porocytes, or pore cells, form the holes in the sponge's body through which water and food particles pass, although the holes in most sponges form between the cells. Myocytes, or contractile cells, surround the openings in a sponge and relax and contract to admit or inhibit the flow of water. Epithelial cells pave the surfaces of the animal. And finally, the hardest working cells of all, the choanocytes, or collar cells, feature minute, moving whips (or flagella) that create the current of water through the sponge's body upon which it depends for its existence. Choanocytes are as important to a sponge as a beating heart is to each of us, but all of its cells do two things that allows us to call sponges animals.

They work together and they communicate with each other.

A sponge's body is held together by a remarkable substance called collagen, a protein found in the bodies of all animals and one of the early clues to the membership of sponges in the animal kingdom. Collagen forms the supportive system that holds the cells together and maintains the specific shape of the sponge species. Among the more obvious truths about animals is that they have to have bodies to keep the cooperating, communicating cells together in one place. Collagen makes

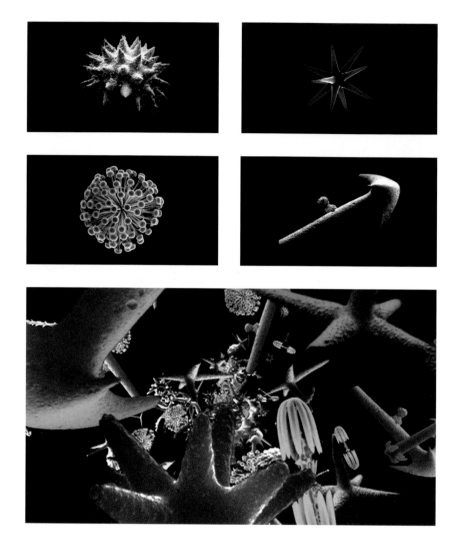

bodies possible. When most people think of a sponge, they picture the bath sponge in the shower stall, but this is really the soft collagen skeleton of a now-departed animal.

Most sponges also have tiny, hard skeletons, called spicules, organized into scaffolding that is the microscopic rival of the Eiffel Tower in Paris, the George Washington Bridge in New York, and other elegant latticework masterpieces so pleasing to the human eye. Like fingerprints, spicules can be used to identify a species, or a group of sponges, not only by type of spicule, but by the way they are arranged in the sponge's body. Our simple, distant relatives actually mine calcium and silicon from sea water to produce these wonderful, glasslike structures, which come in a dazzling array of shapes and sizes. Each begins when spicule-producing cells coordinate in the secretion of a new spicule, a process that continues as the cells divide and migrate to the extremities of the spicule, sometimes growing into a delicate, intricate shape.

In Japan, the crystalline spicule skeleton of a sponge in which repose the hardened remains of a pair of shrimp is a traditional wedding gift, a metaphor in that culture for the eternity of marriage. This little ceremonial sponge is called a Venus flower basket, and as it grows in the sea, weaving its glassy spicules together, it sometimes forms a living refuge for a pair of tiny shrimp. There the shrimp grow, and as they become too large to escape through holes in the spicule web, they are confined forever within the palace of spun glass, a metaphor in our own culture, of course, for a trap. Sponges appear humble in their simplicity, but they have left their mark on Earth on a scale as grand as the Great Wall of China. About 400 million years ago, sponges dominated the oceans as reef builders, a vital ecological chore now handled for the most part by their slightly more complex relatives, the corals (see Chapter 2). Living reefs in the sunlight-rich shallows offer nourishment and refuge to thousands of species, drawn by the easy living with plenty to eat as a

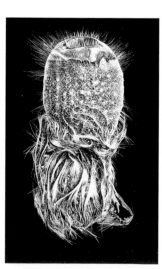

predator, and plenty of places to hide as prey. In the rocks of the Guadalupe Mountains of what we now call Northwestern Texas, you can still see the fossilized remains of a great sponge reef from the Permian epoch, which ended 250 million years ago. The remains of another stupendous sponge reef, much larger than the Great Barrier Reef off modern Australia, covers an arc across most of Northern Europe. The reef was built 200 million years ago by sponges with particularly dense spicule frameworks that fossilized easily into rock so hard it was a favored material for castles and buildings in the Middle Ages.

As kings and queens of ancient reefs, and to a lesser extent today, sponges perform most nobly as spectacularly effective pumps, filtering the water around today's coral-dominant reefs as valuable partners in the ecosystem. Pumping water through their bodies is as elementally spongelike as intelligence is humanlike, a trait so definitive it is impossible to imagine the animal without it. A typical sponge can pump an amount of water equal to its own volume of tissue in about eight seconds, or to express that power in another way, a sponge the size of a human finger can circulate five or six gallons of water through its body in a single day. The entire contents of bays and reefs with large sponge populations flows in and out of these simplest of animals every day, and in the shallows, the outflow from the sponges can be strong enough to disturb the surface of the water.

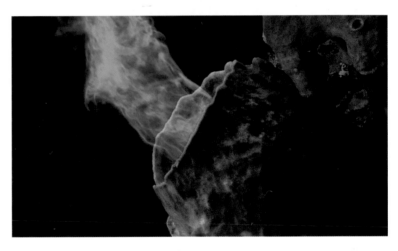

The private life of sponges has captivated enough researchers that we now know their pumping habits and strategies vary from species to species. Most of them pump water with their microscopic whips and some have daily pumping cycles, slowing the flow in the dark by contracting their larger out-current openings, and opening wide during the times of day when the most food will be drifting by. Every so often, some sponges seem to stop pumping for several days. Incredibly, there are sponges that burp, actually reversing the flow of water through their bodies to clear sediment after a storm.

Pumping is not just stage magic for a sponge, but rather the sole means for these animals to

A DIVERSITY OF MICROSCOPIC SPONGE SPICULES (OPPOSITE, TOP), AND THE CRYSTALLINE SKELETON OF A GLASS SPONGE (BOTTOM). ☐ CHRISTINA DIAZ RELEASES FLORESCENT DYE NEAR A SPONGE (THIS PAGE, TOP), AND WATCHES DYE BEING EJECTED WITH THE SPONGE'S CURRENT (BOTTOM).

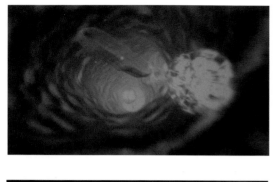

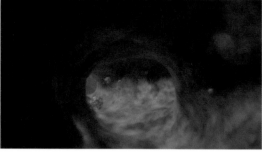

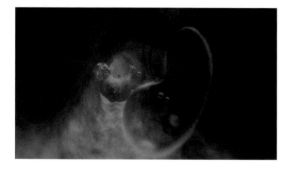

breathe, get a meal and have sex. Imagine, for a moment, that you have shrunk to the size of a single-celled bit of organic matter drifting in the warm sea near a big barrel sponge. You are just hanging there in the water, but gradually you feel a slightly more powerful current that increases as you are drawn nearer to the sponge. Suddenly, you are within the creature, roaring through increasingly smaller canals past cells making new spicules, into chambers lined with thrashing choanocyte whips, and if you have not been snagged as the sponge's next meal, you are eventually swept into the exit cavity, and back into the sea outside. Had Aristotle been able to take this ride into a sponge and seen its workings at the microscopic level, he would never have doubted that this intermediate life form was an animal. A sponge must pump over a ton of water through itself to get an ounce of food. And its way of feeding is closely tied to its unicellular ancestry, in which endless pumping makes food-laden water available to the individual cells. As the minute particles of bacteria and organic debris flow past the whipping flagella of the choanocytes, they are trapped by the amebocytes, engulfed by their soft walls. Inside the cells, the microscopic bits of food are surrounded by a membrane that acts like a minuscule stomach in which secreted enzymes transform the food particle into usable nutritious chemicals for sustaining life. The contents of the tiny stomach are passed on to other cells that, because they have specialized to perform other tasks, are not able to capture a meal. At the same time, oxygen from sea water diffuses directly into the cells, and carbon dioxide is released into the out-flowing current. In other words, the water flowing through a sponge acts as the respiratory system.

A sponge doesn't exactly get in the mood, but at certain times of the year in tropical reefs, the temperature of the water or other sponge sex-triggers transform some of the choanocytes into sperm (notice the resemblance?) and some of the amebocytes into eggs, depending on the sex of the sponge at the time. Again using pumping as its chief ally in the survival game, the sponge blasts its sperm or eggs into the water in a cloud of potential life, and the next generation is on its way. The cliché line in sponge courtship is, "Do you mind if I smoke?" Sponges, like all animals, can only mate with members of their own species, so they coordinate their smoking to ensure their eggs and sperm will meet in the water column. After fertilization, the sponge egg divides repeatedly and arranges itself into a larval stage—in a process

THESE STILLS OF AN ANIMATED RIDE THROUGH A SPONGE (ABOVE) SHOW PROTISTS AND BACTERIA CARRIED IN THE CURRENT. A CELL PULLS APART AS IT PRODUCES A SPICULE (OPPOSITE PAGE, TOP). CHOANOCYTE FLAGELLAE WHIP AROUND INSIDE A CHOANOCYTE CHAMBER, CREATING A CURRENT THROUGH THE SPONGE (BOTTOM).

again like many other animals. During this stage, the flagella face outward to propel them through the water until they settle on the bottom and anchor themselves to reefs, rocks, or some other good home, where they rearrange themselves into adult sponges.

Though sponges appear anchored in place, some of them are mobile. Very patient investigators have discovered that sponges can move by tracing their daily positions on the glass walls of aquariums and observing that they do get around, very slowly of course, propelled by the cumulative motion of their cells detaching from the main body and assembling in a new place, thus moving the edge of their bodies. The ability of sponge cells to move and reform a body is spectacularly documented in a legendary experiment in which a sponge is squeezed through a fine-pored cloth, and eventually gathers itself back into small clusters, some of which will fuse together to build tiny sponges once again. And more profoundly, two different species of dismantled sponges will only re-aggregate with cells and pieces of their own species, demonstrating a powerful animal trait of recognizing 'self' from 'other.' This ability to differentiate between self and other is the beginning of what in more complex animals we call an immune system. An animal body that will accept its own more readily than another greatly reduces the threat of invasion by foreign cells.

The entire repertoire of a sponge—its behavior—is clearly that of an animal. *All* animals exhibit behavior. The elementary cellular form and function of a sponge, though, place it at the base of the evolution of complexity in the animal kingdom. Its possible derivation from a colony of single-celled organisms leads us to believe that this unlikely creature might be the earliest ancestor of all animals, including ourselves. Finally, 2,000 years after Aristotle's early musings, and 150 years after Darwin's brilliant flash of insight about the role of time and natural selection, the confluence of genetic research and fast computers allows us to know for sure.

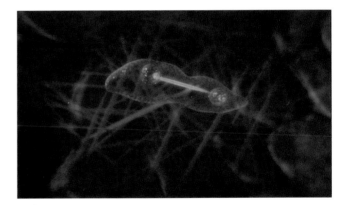

Sponges lie at a critical juncture in the evolution of more complex life forms on this planet. They're clearly basal to other animals. Their common ancestor came from a unicellular world.

Mitch Sogin, Biologist

The definitive proof that a sponge is our most ancient ancestor took some time to find. Until recently, our search for the first animal depended on interpretations of ancient fossils and anatomy. Now, though, we are beginning to decipher those immortal markers we call genes, which contain the evidence of the ancestry we share in common with each other and with the very first animal to have

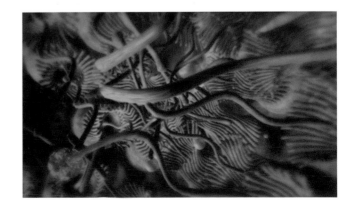

every lived. We have been able to create a new kind of family album, one where the photographs come from genetic material in the nuclei of our cells we call DNA (short for deoxyribonucleic acid), which contains the basic information to build an animal. Instead of great aunts, the keepers of this wonderful new album are people like biologist Mitch Sogin.

As new technology became available, Sogin decided to search the genes of different animals and the possible ancestors of animals among the protists to see if he could find the animal group that gave rise to all the others.

"In looking for the origins of animals, particularly from the perspective of a molecular evolutionist, you can take a top-down approach in which you say, 'I recognize that the first animal certainly was not a cow or a pig or a human. It must have been something much more simple.' And so the top-down approach would be to try and predict which of the various animals that we know about—most likely from marine environments—are those that are likely to be early animals."

In the 1980s, Sogin set himself the challenging task of discovering what creature lay at the base of the animal kingdom by looking where no one had looked before—inside the sponge's genetic code. First, he had to map the DNA inside their cells, a painstaking process known as gene sequencing in which the sequence of nucleotides (represented by letters) in the genetic code is mapped for comparison with those of other animals.

"Genetic sequences are very much like blueprints for constructing an organism," Mitch says in his lab at Woods Hole, Massachusetts. "They define everything that there is to know in order to generate a body plan. You have a set of genetic blueprints that defines who you are. I have a slightly different set of genetic blueprints that defines me. It's very much like the blueprints for building buildings, or cars, or whatever it is you have to have a plan for making. We can compare the genetic blueprints of different things. So, for example, I can compare the genetic blueprint of an automobile to that of a covered wagon. There are a minimal number of elements in those blueprints that are common between those very disparate kinds of vehicles. I can do the same thing with animals."

Since the same genes are inherited in long lineages of organisms through geological time, all animals share many of the same genes. Sogin compared the genetic blueprints of sponges with those of other animals—mammals, insects, worms, and others—by focusing on one of the genes they hold in common. But these genes have changed through time and their subtle differences can tell an evolutionary story. If the sequence of genes of two animals revealed few variations, the animals are closely related. By grouping animals by their shared common sequences, Sogin traced an evolutionary family tree, knowing that the animal at the base of the tree would be our oldest ancestor.

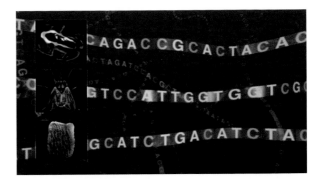

"For a long time, biologists have argued that sponges are basal to all other animals," Sogin says. "Fossil spicules of ancient sponges are found, for instance, in just the right place in time when multicellular animals are thought to have emerged. But there really wasn't any objective, certain way to make that claim until the advent of molecular sequencing capabilities."

"We precisely determined the sequence of a gene from a sponge and compared it with the same gene in a jellyfish," Sogin explains. "Then we compare that same gene in a fly, a fish, a frog, and a human." He focused in on one particular gene to see how it changed and varied in the DNA of the different animals over evolutionary time, and discovered that sponges, indeed, were the most basic, earliest animals that had transformed life on Earth. "The sponge was the first animal with the genetic blueprint for living large," Sogin says. "All animals with more than one cell are based upon that same blueprint."

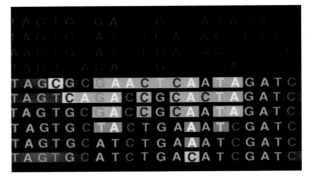

> *Not a single one of our ancestors died in infancy. Not a single one of our ancestors was felled by an enemy, or by a virus, or by a misjudged footstep on a cliff edge, before bringing at least one child into the world. Thousands of our ancestors' contemporaries failed in all these respects, but not a single solitary one of our ancestors failed in any of them.*
>
> Richard Dawkins,
> *River Out of Eden: A Darwinian View of Life*

At the beach on Tampa Bay, the settling heat of late afternoon has driven everyone to hammocks and shadows except for the children who are sprawled on the sand. A mud-colored dog patrols the picnic tables and a few gulls wheel overhead, looking for scraps. An army of ants has formed skirmish lines up the sides of stewing garbage bags, and offshore, fish break the surface as the angle of the sun over the water acts as the cue to begin an evening's feeding in the low light that gives them an advantage over their prey. The family reunion goes on.

MITCH SOGIN (OPPOSITE). ☐ THE SAME GENE FROM DIFFERENT SPECIES IS SEQUENCED (THIS PAGE, TOP). SIMILAR SEGMENTS ARE COMPARED TO ESTABLISH THE PATTERN OF RELATIONSHIP (MIDDLE). THE SPONGE IS THE ANCESTRAL ANIMAL (BOTTOM).

CHAPTER 2

THE HYDRA
AND THE MEDUSA

Life on the Move

Motion is a very beautiful thing. Movement is so much a part of our lives it's like breathing. We don't even think about it.

Jack Costello, Biologist

Take a moment to follow these instructions: Raise your right hand in front of your eyes. Make a fist. Make the peace sign with your first and second fingers. Make a fist again. Open your hand. Read the next paragraph.

What you just did was exhibit a trait we associate with all animals, a trait called, quite simply, movement. And not only did you just move your hand, but you moved it after passing the *idea* of movement through your brain and nerve cells to command the muscles in your hand to obey. To do this, your body needs muscles to move and nerves to activate and coordinate the movement, whether voluntary or involuntary. The bit of business involved in making fists and peace signs is pretty complex behavior, but it pales by comparison with the suites of thought and movement associated with throwing a curve ball, walking, swimming, dancing, landing an airplane, running down prey, or fleeing a predator. But whether by thought or instinct, you and all animals except sponges have the ability to carry out complex sequences of movement called behavior. In fact, movement is such a basic part of being an animal that we tend to define *animalness* as having the ability to move and behave.

Obviously, something changed since sponges led the parade of animal life onto Planet Earth over one-half-billion years ago. The ancient Animal Eve was a barely differentiated collection of cells capable of transmitting messages only at the most primitive chemical level. It had no muscles or nerves. Although some sponges can slowly change location by literally building themselves out into a new direction, they don't do it by transmitting nerve impulses to command movement.

So why, when, and how did animals develop the ability to move?

The little creature whose natural history I am about to present has revealed facts to me which are so unusual and so contrary to the ideas generally held on the nature of animals, that to accept them demands the clearest of proofs.

Abraham Trembley, 1744,
*Mémoires Concerning the Natural History of a Type of
Freshwater Polyp with Arms Shaped Like Horns*

19

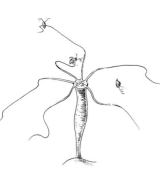

Abraham Trembley was a Swiss scholar and teacher living in Holland in the eighteenth century. His insatiable curiosity about the natural world led him to lay the groundwork for a great leap in our understanding of animalness. To Trembley's contemporaries, a decision about whether something was an animal was elemental not only to their scientific understanding of life on Earth, but to their philosophy as well. The word animal derives from the Greek word for soul—*anima*. Did all animals have souls? And if certain animals they observed were able to regenerate themselves from buds or severed parts, did each new animal have a soul? Was there a soul in the part that regenerated?

And, they wondered, how simple could an animal be?

Trembley was investigating life in a pond near where he lived when he came upon tiny, green, plantlike things that looked for all the world like tufts of grass. "The shape of these polyps, their green color, and their immobility gave one the idea that they were plants," he wrote. "This same initial impression was evoked in many people who saw the polyps for the first time when the creatures were in their most usual position. Some said that the polyps were bits of grass; others compared them to the tufts on dandelion seeds."

Then to his astonishment, he watched them move. "The first movement that I noticed was in their arms, which bent and twisted slowly in all directions. Thinking that they were plants, I could scarcely imagine that this movement was their own," he wrote in his now-legendary *Mémoires Concerning the Natural History of a Type of Freshwater Polyp with Arms Shaped Like Horns*. The title of his memoir would hardly make Oprah's Book Club today, but in it he began to construct a bridge across the abyss that once separated human beings from their understanding of the origins of animal life and behavior.

Trembley was a scientist, and like the modern members of this tribe of discoverers, he then made systematic observations to support his hypothesis that these freshwater polyps from his pond, now called hydra, were animals instead of plants because they moved and caught and ate prey. He

took some of his subjects into his lab in a jar of water, offered them food, and watched them eat. He also placed a jar of them on the windowsill and noted that they moved slowly toward the lighted side. After weeks of painstaking study, he decided correctly that they were animals instead of plants, and the key to his conclusions was that they moved and ate. "I saw them digest animals as long and even longer than themselves. Thus I had hardly any further reason to doubt that they were carnivorous animals," he noted.

Two hundred and fifty years later, Trembley is cele-

HYDRA CATCHING A WATER FLEA (THIS PAGE, TOP). WOODCUT FROM TREMBLEY'S BOOK SHOWING HIM AT WORK WITH HIS STUDENTS (BOTTOM). □ A JELLYFISH (OPPOSITE PAGE, TOP) AND SOLITARY CORALS (BOTTOM) REPRESENT THE TWO BASIC CNIDARIAN BODY FORMS.

brated not only for his insights into his miniscule carnivores but as nothing less than the founder of experimental biology itself. "The birth certificate (of experimental biology) was signed by the Swiss naturalist Abraham Trembley, who with his work on freshwater hydra began the line of those naturalists who, starting from being passive observers of nature, became active experimenters," writes science historian Joseph Schiller.

Abraham Trembley's discovery that the green things in his pond were active, predatory animals was a revelation for eighteenth-century science, but in the centuries that followed it would lead to even more profound conclusions about animal life: Their earliest ancestors, we would learn, invented nerves and muscles. The little creatures in his pond were polyps, members of the group of animal life known as cnidarians (ny-dair-i-ans) that includes not only his freshwater creatures, but the much more common saltwater polyps, including corals and sea anemones. In addition, there is the flip side of the cnidarian family—the medusae—most familiar to us as jellyfish.

Ancient cnidarians were pioneers of the complexity we now take for granted. Trembley's discovery began a line of inquiry that eventually brought us to the understanding that the polyp and medusa not only move, but they pioneered the use of the muscles and nerves that are so much a part of what we consider animalness. All animals except sponges have nerves and muscles, and since cnidarians are the simplest animals to possess this complexity, their earliest ancestors were very likely the first.

Cnidarians are also the first animals with an actual body of definite form and shape. The presence of tissue allowed a body to take on a shape, a level of organization in which groups of cells of the same kind bond together to collaborate in performing the same functional chore. The invention of tissue allowed skin, muscles, and nerves to function as structures, which are much more effective than separate cells. This body also had the first mouth and a digestive cavity, which would eventually allow cnidarians to become voracious predators. These cnidarian innovations, so radical and important to survival, were passed down with increasing refinement to all other animals as they evolved through time.

For me the really fascinating questions in biology are all about origins. Where have we come from? What sort of things started off the evolution of behavior? And you gotta go back to a group like the cnidarians in order to study this question.

Ian Lawn, Biologist

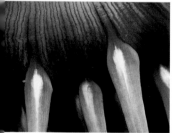

According to the fossil record, the earliest cnidarians made their appearance about 600 million years ago, most likely as fleshy polyps anchored to the seafloor in a world where no animal moved, swam or scuttled. Although no fossils of intermediate animals have been found, the earliest cnidarians probably evolved from sponges, in response to the constant pressures of procreation and survival. They developed additional specialized cells that included nerves and muscles. The reason we have so little evidence of early cnidarians in the fossil record is that the bodies of these simple creatures consist mostly of water and jelly, leaving few enduring traces.

This amazing body plan evolved into two distinct forms. Although they use the same functional structures, the body plans are essentially mirrors of each other. By turning the same shape upside down, cnidarians devised two distinct ways of living: the polyps and medusae. The polyps include Trembley's little pond critters, anemones, and soft and hard corals. They feature cylindrical bodies with one end attached to the seafloor, a piling, rock or other surface, while the other end freely waves a tentacle-ringed mouth in the water. The medusae are their mirrors, mostly bell shaped but

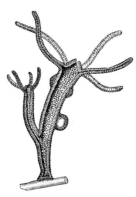

some shaped like cubes, with tentacles and mouths facing down. (The down-hanging tentacles resemble the snakes of Medusa's hair from Greek mythology, hence their name.)

About 10,000 species of cnidarians populate the seas, lakes and rivers of Earth, although most are marine. Trembley's freshwater polyps are a relatively minor branch of the group, and the more familiar jellyfish, anemones, and corals are by far the most common. Whether polyp or medusa, they all share the same basic architecture for living. Some are microscopic, some as large as nine feet in diameter, but they are all radially symmetrical, with nerve nets for transmitting impulses in response to stimuli, a single body cavity, and an opening in the body that serves as both mouth and anus. This shape means that a cnidarian's body does not have a right or a left, but is the same all around, which is a smart way to live if you are attached to the bottom or floating in the three-dimensional ocean waiting for a meal to drift by.

They are creatures so simple that scientists once considered them plants. But they're the critical group to study if you want to understand motion and behavior.

Jack Costello, Biologist

Ian Lawn, Jack Costello, and most other biologists believe the appearance of early cnidarians about 600 million years ago revolutionized animal life, but these scientists don't have to be time travelers to study them. A common ancestor of the ancient hydra and medusae passed on their inventions to the rest of the animal kingdom, including our vertebrate ancestors, but the basic cnidarian body plan, in both its forms, continues to exist.

Although they are predominately water, cnidarian bodies are quite a bit more complicated than sponges. The body wall of a polyp or a medusa has two layers of cells, one specialized for protection that covers the outside, and an inner layer lining a stomach cavity for digestion. All cnidarians are carnivores. Meat-eaters. Digestion begins in the central cavity, a primitive kind of stomach, and is completed within specialized cells. Whiplike flagella on the cells lining the stomach cavity keep

THE FLESHY BODY PARTS OF ANEMONES (OPPOSITE PAGE). □ A CROSS-SECTION OF A HYDRA SHOWS THE INTERNAL STOMACH CAVITY (TOP RIGHT). JELLYFISH TRAIL LONG TENTACLES (RIGHT).

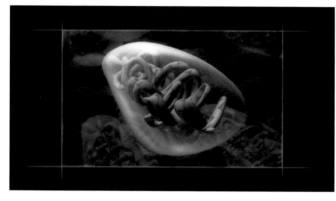

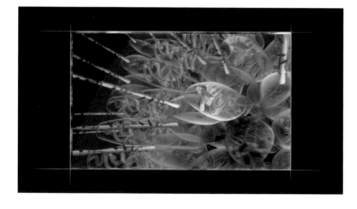

observed them at work, firing out of the capsule sac with the acceleration of a rifle bullet at nearly 40,000 times the acceleration of gravity, everting a hollow filament through which poison is injected when the harpoon hits its target.

After the nematocyst silos are empty, the nematocyst-growing cells, called cnidoblasts, have to grow new ones before the next attack. And because each nematocyst can fire only once, it must not do so indiscriminately. If a jelly brushes against the glass wall of an aquarium, for instance, its nematocysts will not fire. If it touches a potential meal or enemy, only the nematocysts in the immediate vicinity of the stimulus will fire. There is little evidence that the stimulus signal is passed to other parts of the body, which means that each nematocyst cell is independent, containing within itself the sensory properties required to fire. Most likely, the nematocyst cell is able to chemically differentiate between the animate and inanimate stimulator, a very sophisticated bit of work for animals that look pretty much like The Blob.

Nematocysts are among the most complicated examples of a cell product in the animal kingdom. Without nematocysts, cnidarians would lead very different lives. The Defense Department spends billions of dollars trying to develop weapons as effective as nematocysts, but those brainless wonders did it many times over to adapt to the characteristics of their targets. The harpoons of some nematocysts are perfect for piercing fish scales, some for combat with other cnidarians in turf battle, some for penetrating soft flesh. The toxin they inject can be powerful enough to kill a human being. The cubomedusae, or box jellies, of the waters off northeast Australia claim twice as many human victims as sharks. On the other end of the scale, the nematocysts of most of the anemones we touch in tide pools are so small we barely feel their effect, only a kind of tingling stickiness. If you touch an anemone with your tongue, though, you'll definitely get stung because the softer tissue of your tongue allows the harpoon to penetrate. (*Do not try this.*) While nematocysts are chiefly weapons,

ANIMATED STILLS OF A BATTERY OF UNDISCHARGED NEMATOCYSTS (THIS PAGE, TOP). SINGLE UNDISCHARGED NEMATOCYST (MIDDLE). NEMATOCYSTS WITH THEIR BARBED HARPOONS DISCHARGED (BOTTOM). ☐ FIGHTING ANEMONES INFLATE SPECIAL TENTACLES LADEN WITH NEMATOCYSTS (OPPOSITE PAGE).

cnidarians have adapted them to other uses as well. Hydra use them to attach themselves while moving. Some anemones use them to build tubes in which to live. Others, like siphonophores, dangle lures loaded with nematocysts to attract their prey.

As bellicose as all that harpooning and poisoning sounds, cnidarians can be among the most cooperative animals on Earth. One of the most dramatic examples of their cooperation, Australia's Great Barrier Reef, was created over millenia by corals forming huge reefs that can be seen from orbiting spacecraft. The little coral polyps reproduce by dividing in half over and over to create new members of the colony, remaining attached to their offspring while building the structures we call reefs. Just one of these coral colonies can contain millions and millions of individual animals, joined together for life. The intricate surfaces of a reef are covered with living corals, built upon the hard bodies of their departed brethren who have turned into limestone.

And not only do the hard corals of a reef work together to increase their chances for survival, but within each tiny polyp is another example of cooperation. These corals play host to symbionts, called zooxanthellae, which are tiny photosynthetic algae. By teaming up with zooxanthellae, corals guarantee themselves a supply of oxygen and nutrient-rich by-products of the algae to supplement the rest of their diet of drifting zooplankton captured by their tiny, nematocyst-studded tentacles. The corals, in turn, offer these precious solar collectors nitrogen, carbon dioxide, and a sheltered place to live within their colossal reef castles.

Love among the cnidarians is also a cooperative affair and about as diverse as you can get. Some species of the medusa side of the family are hermaphroditic, but most have separate sexes. Sperm and eggs are cast into the water, and the lucky ones get together. Some polyps, such as corals, also spawn by broadcasting sperm and egg, and their sex act is orchestrated by moonlight or other environmental cues. Most tiny hydra produce daughter polyps, which simply pop off from their parents, especially when times are good and the water's full of food. Some anemones reproduce by

cloning themselves, asexually dividing into exact copies of themselves and forming dense colonies to control large areas of turf on the seafloor. Clone wars occur between different colonies from time to time, when an anemone from one colony tries to encroach upon another's territory. For these battles, the anemones use specialized fighting tentacles spiked with nematocysts. The conflict can go on for days before one has been stung enough times to force retreat.

As is always the case when investigating mysteries through the lens of deep time, there is still debate over which came first, the polyp or the medusa. Most bets are on the polyp. And then,

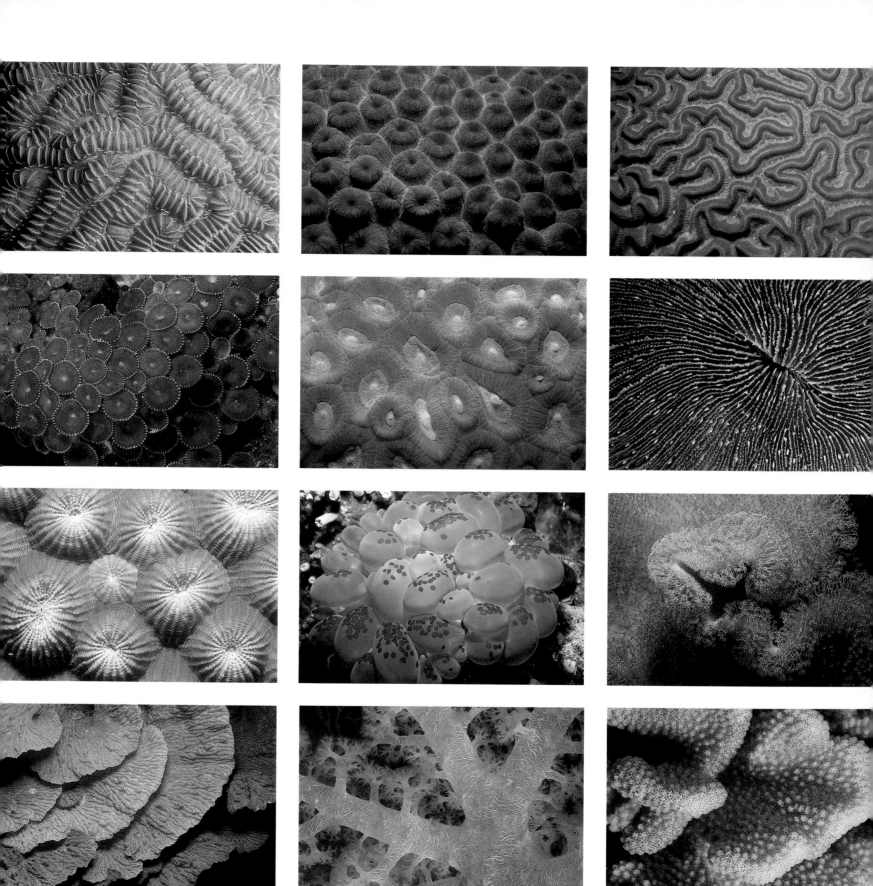

although no one is sure exactly when or how it happened, some of these anchored cnidarians took off in a dramatic new direction. Equipped with primitive nerve nets and muscles that could flex in response to stimuli, and urged by the power of natural selection, they began a transformation. The edges of their mouths extended and developed armlike feeding structures. The polyps' food-grabbing tentacles became long, thin strands, the cylindrical stalks of their bodies became gelatinous bells, and as they began to pulse, they left the seafloor and drifted away. The jellyfish had arrived on Earth. The medusae may have been among the first animals to actually swim the world's oceans, where they remain today as dominant predators in the food web.

Because cnidarians are still around, we can watch a wondrous echo of that moment of transition from polyp to medusa that happened so long ago. The polyps that demonstrate the transformation, in a process called strobilation, need special laboratory equipment to be seen in action. They are called moon jellies, and like many of the medusae, begin their lives as tiny polyps, anchored to a rock or other solid surface on the ocean floor. A few times every year, triggered by the rhythms of the sea, they undergo a transformation that very likely mimics the emergence of the first medusae over one-half-billion years ago. The polyps are as different from adult moon jellies as caterpillars are from butterflies, so different and so small that the polyps of some species of jellies have yet to be discovered.

"It's a phenomenal process. At some point, those polyps begin to divide," says marine biologist Jack Costello, "and it's almost like they form little plates, one on top of each other." Each polyp forms dozens of orange-colored plates, and each plate pulses like a baby bird trying out its wings at the edge of its nest until in turn it breaks off and becomes a single animal, called an ephyra. Over the course of a year, the tiny juvenile jellyfish feed on plankton as they develop into adults and can reach as large as two feet. Some say watching this happen is like riding a time machine into the deep past to the moment when polyps first rose from the seafloor to become the swarms of drifting predators that we know today.

The process begins anew when the adult moon jellies gather in massive swarms in coastal shallows. The males cast threads of sperm into the water; the females collect the threads with their frilly arms and ingest them, fertilizing their eggs. The eggs develop into simple balls of cells that swim free from the mother medusa to settle on the seafloor and grow into a tiny polyp. Within days, the bodies of some of the mother jellies begin to break down and disappear, but their offspring will live for years, budding tiny jellies to populate the oceans.

The life cycle of moon jellies is natural magic, but the swimming behavior of the adult jellies also intrigues Jack Costello. He enters their watery world with diving gear and brings back their stories with a video camera, capturing hours and hours of their behavior. They seem to swim almost constantly, but they go nowhere. The jellies expend enormous amounts of energy to move. So what do they get from moving? "They do spend all their time swimming, and they really don't make

A DIVERSITY OF TROPICAL CORALS AND THEIR KIN.

much forward progress," Costello says. "That leaves us asking, Why would they spend their time swimming?"

This is a classic statement of a puzzle of form and function: Does the jelly's form serve a function that is not immediately obvious? All animals balance their food and energy budgets to produce a net gain and thus grow larger, getting stronger or faster, and, on the bottom line, reproducing and surviving. Obviously, jellyfish are survivors.

But Costello was confounded because the body of the moon jelly did not seem to be designed for easy movement through the water. In fact, its body plan seems shaped to slow it down. "That round disk shape is probably one of the least effective shapes for forward progress that we can imagine. A flattened shape moving through the water presents the most resistance, in what we call drag, of any form you can imagine," Costello says.

In a highly refined echo of the methods of observation pioneered by Abraham Trembley, Costello's observations depend on systematic inquiry to either support or disprove an hypothesis, the self-correcting process known as the scientific method. And we have come a long way from experiments conducted in jars of water sitting on a windowsill in the sun. Instead, in a modern laboratory, we can record and analyze the cnidarians in carefully controlled tanks that mimic their life in the wild. "What we wanted to do was look at the action of swimming and look at how the jelly interacts with the fluid around it." Costello's subjects this time were the size of water drops. Into the tank with the jellies, he added tiny, visible particles, which allowed him to actually watch and videotape the flow of water around the animals. As the jellies swam, they created a visible current, and it was in their creation of that current

that he found the solution to his form-and-function puzzle. "What we found looking at it is that this very high-drag body form is very good at creating vortices and flow around the bell margin," he says. "It's the flow created by swimming that is bringing all the prey into the capture surfaces. The body we think of as bad or ineffective for forward motion is very effective for creating the flow which enables the

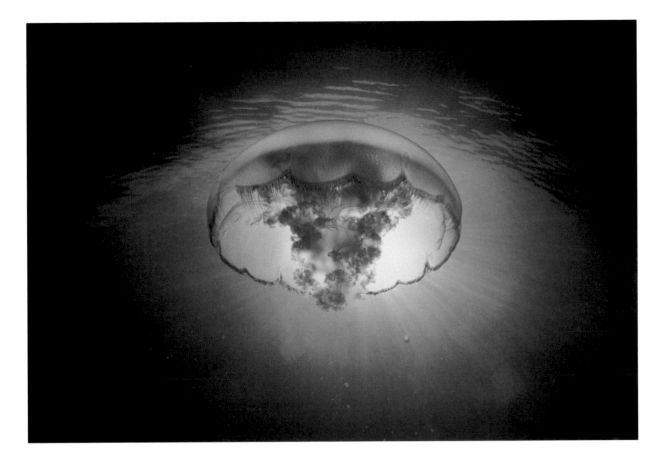

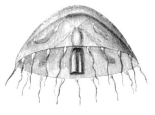

animal to feed." As the jelly's bell pulses, the water flows in under the bell margin through the tentacles, bringing food particles in contact with both the tentacles and the oral arms, where nematocysts can shoot their harpoons to capture the prey.

Figuring out what a jelly gets from a round body that doesn't get anywhere provides insight into their origins, too. The first animal to launch itself from the seafloor—perhaps at the budding end of a creature like a moon jelly polyp—may have done so to ensnare prey, not to give chase. "It's an incredibly simple but very effective system," Costello says. "It becomes a little more sinister when you realize that it is the mechanism by which these animals move through the water and essentially kill all their prey." The size and shape of the bell of each species of jellyfish determines the flow pattern around the animal, which means that each mines a particular niche in the ocean food web.

TINY MOON JELLY POLYPS (OPPOSITE PAGE, TOP). A SINGLE EPHYRA IS RELEASED (MIDDLE). A SWARM OF EPHYRA AND JUVENILE JELLIES (BOTTOM LEFT). MARINE SCIENTIST JACK COSTELLO (BOTTOM RIGHT). □ ADULT MOON JELLY (ABOVE).

Because of their incredibly simple but very effective system for living in the sea, cnidarians have flourished as predatory carnivores. In today's oceans, these durable, successful animals mark time in centuries and have assumed forms that take advantage of an enormous spectrum of ecosystems, from the patient anemones of the tide pools to the giant jellies of the open ocean depths. Some are utterly startling. One polyp, the anemone called *Stomphia*, looks vulnerable as it sits anchored to a rock on the ocean floor, but it defends itself by detaching from the rock and swimming to safety. Actually swimming.

The longest animal in the world, even longer than a blue whale, is *Praya*, a combination of the two basic cnidarian shapes, the pulsing bell of the medusa and a long train of modified polyps with tentacles. A single colony of these animals may be 120 feet long. Strung like a giant fishing net through the water, *Praya* feeds itself by chancing into prey and stinging them to death. Another medusa, *Colobonema*, has evolved a surprising line of defense. When startled, it detaches its tentacles and leaves them behind as decoys to distract would-be attackers, much in the way a jet bomber dispenses chaff to confuse incoming missiles.

All of their remarkable behavior and success for more than one-half-billion years, though, rises from the cnidarian's ability to move. Yours, too.

Take a moment to follow these instructions: Raise your right hand in front of your eyes. Make a fist. Make the peace sign with your first and second fingers. Make a fist again. Open your hand. Say, "Thanks, Cnidies."

A DIVERSITY OF TENTACLES AND BELL SHAPES ARE EVIDENCE OF THE MANY WAYS JELLY FISHES MAKE A LIVING.

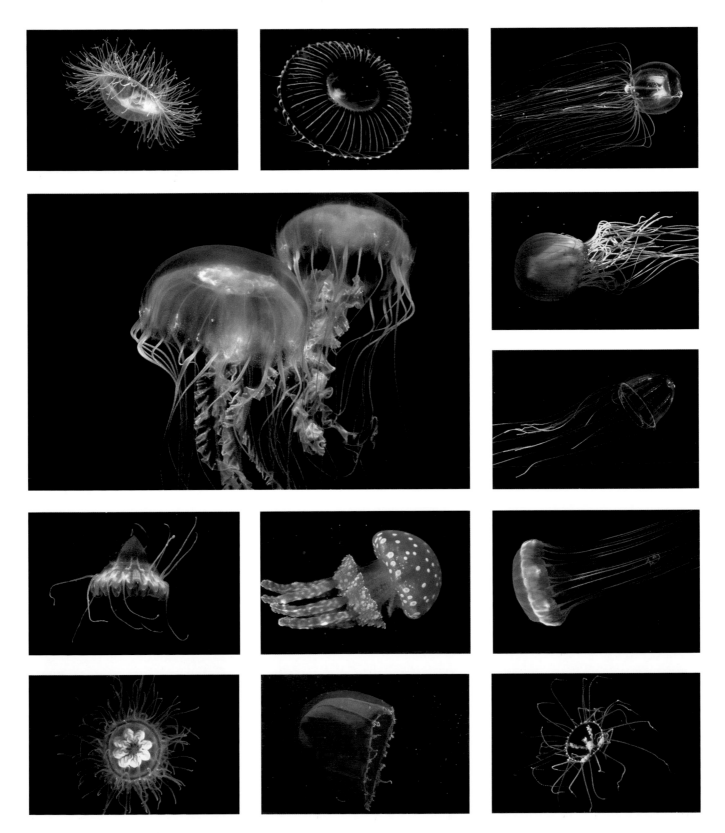

CHAPTER 3

GETTING A HEAD

The First Hunter

When you collect marine animals there are certain flatworms so delicate that they are almost impossible to capture whole, for they break and tatter under the touch. You must let them ooze and crawl of their own will onto a knife blade and then lift them gently into your bottle of sea water.

John Steinbeck, *Cannery Row*

And now, the news . . .

FLATWORMS ATTACK 200 MILLION PEOPLE

(Geneva, Switzerland) *The World Health Organization reports that more than 200 million people worldwide are carrying parasitic flatworms and over 500 million people are at risk in 74 countries. The resulting disease kills up to 20,000 people a year and leaves other victims in a state of chronic ill health. It's known as intestinal schistosomiasis, and is caused by the tiny* Schistosoma *spp. flatworms, also known as blood or bladder flukes.*

Schistosomiasis has been recognized since the time of the Egyptian pharaohs, and the worms that cause it were first discovered in 1851 in Cairo by a young German pathologist, Theodor Bilharz. The disease is also called bilharziasis, in his honor.

THIRTY-SEVEN-FOOT TAPEWORM SETS RECORD

(Great Grits, Mississippi) *On September 5, 1991, doctors extracted 37 continuous feet of tapeworm from Sally Mae Wallace of Great Grits, Mississippi, setting a new record for the United States. Tapeworms are segmented flatworms, which attach themselves to the intestines of their hosts. The chief symptom of tapeworm infestation is weight loss, and tapeworm eggs have been put in capsules and sold as diet pills. When the desired amount of weight had been lost, you take another pill to kill the tapeworm. The largest tapeworm ever found measured over 90 feet in length and had over 2,000 segments.*

"After about 20 feet of that thing had come out," Sally said, "I just knew I had the record. I was really filled with joy."

CARNIVOROUS WORMS ROAM WILD IN THE UK

(Dunoon, Scotland) *An invasion of alien predators in the Scottish Highlands is threatening to destroy thousands of acres of farms that are home to sheep, cattle and people. New Zealand flatworms, voracious carnivores believed to have arrived as stowaways in potted plants, are eating up the earthworms upon which the highland fields and meadows depend for aeration, drainage, and fertilization. Now, many of the fields are becoming swampy and useless for grazing or planting. Researchers say there is evidence that the flatworm invasion is spreading throughout the United Kingdom.*

"The first time I remember being aware of them was when fishermen came here a few years ago to look for worms for fishing and they complained there were no earthworms," said farmer Tom Hills. "So then they started to look for reasons why there were no earthworms and they came up with the flatworm."

"They're alien," said Mr. Hills' wife. "And they're quite sinister, really. And underneath the ground, you can't see them. But they're devouring earthworms." According to Hugh Jones of the National Flatworm Survey, a couple of dozen flatworms in a single field can eat a thousand earthworms a year. "Initially, they were just regarded as a curiosity, another one of these imported animals that probably doesn't do much harm. Turned out these things are carnivores, they're hunting other animals."

Over the millions of years that animals have been living on Earth, they have learned to exploit every available source of food and to do that, many have become hunters of other animals. At some point in the distant past, the first animals capable of actively hunting showed up with bodies suited for the job, and lions, tigers, sharks, people and all the rest of the world's hunters inherited their tools. The earthworm killers of Scotland, giant tapeworms in Mississippi, and parasitic flatworms that wreck hundreds of thousands of human lives are the descendants of those first hunters and have carried their body architecture into the present. They don't make very good company, but they have a heck of a story to tell.

To hear that story, though, we have to listen very carefully, because the first hunters barely whisper to us from their beginnings over one-half-billion years ago. And we need translators who understand the languages of the past that resonate in fossils of ancient animals, and more recently, the languages of the genetic code, another tracing carried in all living things. One of those translators, Whitey Hagadorn, is a paleontologist—literally "one who knows of past times." He likes the outdoors, spent a lot of time at a university learning the languages of the past, and now burns with curiosity

A NEW ZEALAND FLATWORM ATTACKING AN EARTHWORM (ABOVE). □ WHITEY HAGADORN (OPPOSITE PAGE) EXAMINING A FOSSIL BED IN THE WHITE MOUNTAINS.

about his particular passion, ancient worms. "I'm playing detective in deep time," he says. "I mean, I'm trying to think what was on the seafloor 565 million years ago."

A fossil is any evidence of ancient life—usually bones, teeth, shells, tissue, tracks, or traces—buried and undisturbed in sediment that became rock. Becoming a fossil, though, isn't just a matter of dying and laying around for millions of years. When most animals die, they get eaten or just rot in a process that depends on oxygen to break down their body chemistry into carbon and other elements, leaving behind hard bone, teeth or shells. If these hard parts get buried by sediments that turn to rock, they become fossils. On rare occasions when an animal dies and is quickly covered up by sediment or ash, the oxygen can't get to its body and then the eons of pressure transforms its entire body into a fossil. It's the exquisite details of anatomy left by these soft-body fossils that often provide major breakthroughs in the study of paleontology. Only in rare cases have sponges, cnidarians, and flatworms spoken to us from deep time through their fossils. Their bodies, made of water, jelly, collagen and other goo, don't turn to rock like the bones, shells, and other hard parts of bugs, fishes, dinosaurs and other animals. Still, paleontologists have found fragile traces of worms and other soft-bodied animals, and have discovered spectacular evidence of their presence—their tracks.

Sponges don't get around much, and even though jellyfish, anemones and the other cnidarians had muscles and nervous systems, they couldn't go out and pursue prey. The earliest hunters, though, threw a new wrinkle into the business of living and our ability, now, to interpret their presence on Earth: They moved around and left trails. "The trick that we have as paleontologists," says Whitey Hagadorn, "is to go back in the geologic record and look in the rocks for evidence of tracks or trails

of ancient organisms, particularly for intervals of time when maybe the animals themselves aren't preserved. In a way, an animal's tracks record its movement, its behavior, its size, writing its steps in the earth."

> *Fossils turn up in the darndest places.*
> Niles Eldridge,
> *Life Pulse: Episodes from*
> *the Story of the Fossil Record*

In the Inyo Mountains of California, among many other places in the world, Whitey Hagadorn found the tracks of some of the earliest creatures that moved around in just the right place to leave the evidence. The traces look like the remains of an angel hair pasta lunch that had been wrapped in newspaper, a thin impression of a sinuous strand of something that left a micro-trench in the surface

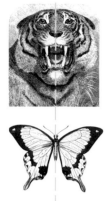

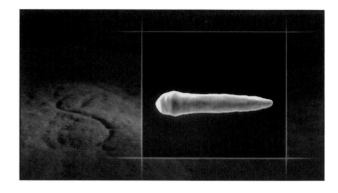

of the rock strata. "I get a clue here, a clue there, and I try to piece them together and create a story to find out who the culprit was. Right about this time, about 565 million years ago," Hagadorn says of those rocks that are now part of California, "there was sort of a revolution in animal body plans. Organisms for the first time became able to move on their own accord in a directed fashion. We know this because we can look at their trace fossils. By looking at this trail, we can certainly say that whatever made it had the ability to move sediment. Prior to this time, there weren't many things that could do that."

The question is: What kind of animal was it?

There are many clues in Whitey Hagadorn's track in the rocks that sketch a portrait of the animal that left them. The trackway is only a few millimeters wide, so the animal had to be thin. The track is rounded, so the creature was probably cylindrical, and there are no scratches along the edge of the trail, so it is very unlikely that the animal had arms, legs or any other appendages. In short, we're talking about a worm, but one other thing about the trackway hints at a revolution in the body plan of animals. The trail shows evidence of a sense of direction. The animal was going *somewhere* and not just drifting in the water. These whispers from deep time carry a profound message: This animal had a head and knew where it was going.

> *Orientation directly towards a goal can be achieved only in three ways. Either the sense organs must themselves be directional . . . or the animal must compare the signals from pairs of similar sense organs on two sides of the body . . . or the animal must be able to compare signals in time.*
>
> Martin Wells,
> *Lower Animals*

Most of the familiar creatures we think of as animals all share certain features, which include a head. So what's the big deal about finding the 565 million-year-old trail of a creature with a head? For starters, something had to come first. And that animal pioneered a way of living that depended upon mobility and hunting, which required a head that contained sense organs of some kind so it

MOST FAMILIAR ANIMALS HAVE A HEAD AND ARE BILATERAL (THIS PAGE, LEFT). EARLY FOSSILIZED TRACK AND THE HYPOTHETICAL ANIMAL THAT MADE IT (ABOVE). ☐ MODERN FLATWORMS' THIN BODIES HAVE COMPLEX ORGANS (OPPOSITE PAGE).

would know where it was going. This breakthrough in design was enormous, and gave those creatures with heads a huge advantage over animals that could only sit and wait for food to float to them, or pulse aimlessly through the water in search of a meal. When animals with heads appeared in the sea, they soon became the dominant form of life on Earth. Imagine your feet being glued to the sidewalk while large, stalking cats patrol your neighborhood looking for a nosh. In the coming eons, animals as small as flies and as large as elephants would evolve using the same basic architecture and strategies for survival as the earliest worms.

The trackways and the bodies of the creatures that left them also reveal another masterpiece of body architecture: They were bilateral. That means that their two sides mirrored each other, and because they had heads, they also had tails. Just like us. We take for granted this basic formula for building an animal body, but the Neil Armstrongs of heads and bilateral symmetry first left their footprints on Earth way back when those rocks began their journey into the present, carrying their messages from the distant past.

> It takes more than a lively imagination to identify with an ameba, which has no fixed front or rear, or with a medusa, which has multiple eyes encircling an umbrella-shaped body . . . Now, we come at last to some invertebrates in which it is easy to tell head from tail, back from belly, and right side from left.
>
> Vicki Pearse, John Pearse,
> Mildred Buchsbaum, Ralph Buchsbaum,
> *Living Invertebrates*

We may never know for sure which specific animal was the very first hunter with a head and all the other equipment for leading that revolution. From the clues in the rocks, we do know it was a creature with features similar to some of the flatworms alive today. About 20,000 kinds of flatworms are alive and well today, in fresh and salt water and other nice damp niches, including the insides of other animals. Their name, *Platyhelminthes* (plat-y-hel-min-thes), comes from the Greek and means, simply, "flat worm."

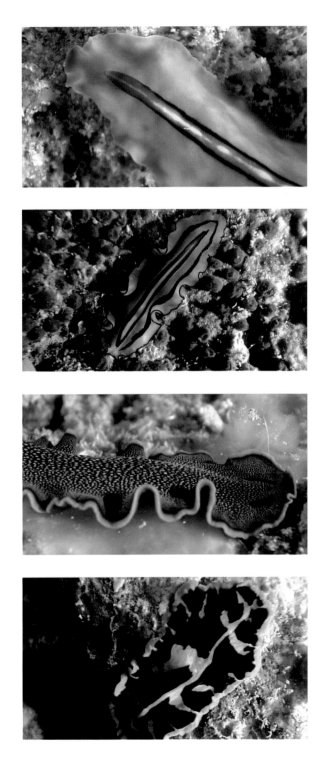

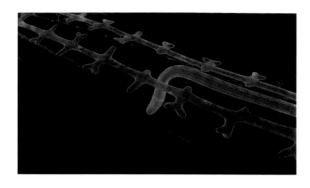

Like their ancient cousins, modern flatworms are soft-bodied, ribbon-shaped animals made of tissue that is further organized into systems of organs. Their bodies have no real central cavity, but rather a layer of tissue in which their organs, nerves, and sensors are embedded. Their bodies have only a single opening, which serves as both a mouth and for waste disposal. Many flatworms eat by extending a tube called a pharynx out of the opening and into their prey. The lining of the pharynx secretes enzymes, which soften the bits of food that the muscles of the pharynx then pass into the flatworm's body. A flatworm distributes its meal through its body in a gut that is not a central channel but more of a network of branches that supplies the food directly to the rest of the body. The system also works in reverse to get rid of waste.

An ancient flatworm ancestor was also the first creature to have a brain—a bundle of nerves to which nerve impulses are transmitted from sensors such as eyes and smell and taste receptors so the animal knows where it is, where its food is, and where its enemy is. The brain interprets the signals from the sensors and the worm can adjust its course, speed up, or slow down. This whole set-up is called a central nervous system and is very big news on the evolutionary front, since all of the things we call animals that were around before ancient worms were much less complex. And they were much less capable of moving around and hunting for food.

The combination of bilateral symmetry and a central nervous system is very logical and works like this: Free-living flatworms (many are parasitic and just ride along in a host) use their network of sensors to detect food, chemicals that lead them to prey or away from predators, objects that can

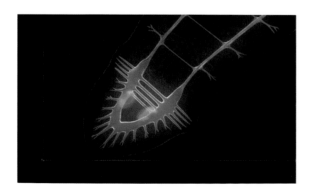

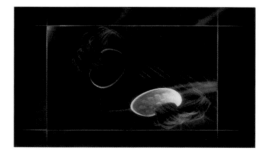

stop their progress, or even water currents. These sensors are paired and placed on opposite sides of their bodies, so the worms can compare the relative position of the sensory input. Two light receptors that are really

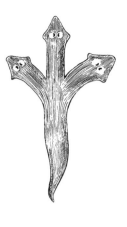

primitive eyes give them stereo sight. Their eyes are tiny, black-pigmented bowls filled with specialized sensory cells, each of which ends in a nerve fiber connected to the brain. Because the black pigment provides shading, light from only one direction hits the sensory cells, causing a chemical reaction that triggers the nerve fiber to send a signal to the brain.

These primitive eyes tell the direction of the light by which cup receives a stimulus and exactly which part of the pigment cup receives the stimulus. If the light is entirely from the right side, only the right cup gets stimulated because the left cup is shaded. If the light is exactly in front, both sides get stimulated in the same part of the cup. If the worm wants to go toward the light, it turns until the same spot in each cup is stimulated by the light. This will mean that the light is directly ahead. While most of these light-sensing cells are clustered in the pair of eyes, some are scattered in different—but bilateral—locations around the body. The other types of sensors for smell, touch and taste are also paired. A flatworm becomes more sensitive—and quicker to react—as it gets closer to the source of the stimulus because its relative intensity becomes much more apparent to two receptors spaced only a flatworm's-width apart.

Flatworms have been trained to run mazes according to light cues and remember where and in which direction to turn or go straight after a few tries, a stupid pet trick that has never appeared on David Letterman but probably should. And for an encore, Dave can cut one flatworm that has learned to run the maze into two, each half will grow a new head or tail depending on which it needs, and both will be able to run the same maze better than an untrained one, quite a feat for a tiny worm.

All animals can regenerate their tissues to some degree, but flatworms are grand champions. We humans and other mammals do it, of course, when we slough off millions of skin cells every day and replace them with new ones. We can also grow new skin, nerves, and blood vessels to cover a wound, or knit our bones together after a break. Other animals can replace arms, legs, and tails, but the general rule is that the more complex an animal is, the less its capacity to regenerate its body parts. Sponges and cnidarians do a pretty good job with regeneration, and can grow whole new animals from tiny flakes or buds of themselves. But their bodies are much simpler collections of cooperating cells than those of flatworms. What makes the regenerative ability of flatworms so interesting and remarkable is that their bodies feature the earliest hints of true animal complexity—brains, organs, a nervous system, heads, and tails. They *look* like animals, and their regenerative repertoire can be startling. Almost every biology student has seen or heard of a multiheaded monster planarian flatworm

A PLANARIAN'S PHARYNX AND BRANCHING GUT (OPPOSITE PAGE, TOP). THE PHARYNX EXTENDED AND FEEDING (MIDDLE). NERVES IN THE HEAD FORM A SIMPLE BRAIN (BOTTOM). STEREO EYES SENSE DIRECTION OF LIGHT (RIGHT). □ A MULTIHEADED PLANARIAN (THIS PAGE, TOP). FLATWORMS HAVE BOTH MALE AND FEMALE SEX ORGANS (BOTTOM).

made by making many cuts in its head and allowing each to regenerate into a whole new one.

Flatworms, like the ones terrorizing earthworms in the muck of Scottish farms, are as effective as lions chasing down gazelles, and because they can regenerate so efficiently, they are very hard to kill.

"How do you kill them?" asks one of the members of the anti-flatworm squad in the Highlands. "I have absolutely no idea because you find them under the stones or rocks and they're just flat. And you stand on them, you jump on them, makes no difference. They don't go away. What can you do? Pulverize them? Turn them into mush?"

The regeneration cells are called *neoblasts*, which means "cells of newness."

When a flatworm is wounded or cut in pieces, these special cells immediately begin to migrate toward the site of the damage. The movement of the neoblasts is either triggered by the release of some chemical from the wound, or by some impulse transmitted in the nervous system, or both. Once the neoblasts get to the site, they contain the genetic information to either produce or induce all of the necessary cells to make the animal whole again.

Obviously, replicating oneself through regeneration is a form of reproduction, and some flatworms depend heavily on it for keeping the generational train on the track. Most, though, are pure sex machines with both male and female organs and twice the chance of getting a date when the time for reproduction rolls around. A typical flatworm, say a half-inch-long, common little number called a planarian, has a pair of ovaries set right behind its eyes, and the ovaries secrete eggs through oviducts to a system of yolk glands that are strung out along each side of its body. The planarian also has a whole series of testes connected by ducts set at intervals along two channels that run from head to tail. Most flatworms have a penis, or several, and some have one or more genital pores for receiving a unique, two-tailed sperm delivered during copulation.

You'd think that with that kind of sexual plumbing, a flatworm would never have to leave the house to get lucky, but most do not self-fertilize. Instead, a flatworm only has to find another flatworm—any other flatworm of the same species. In some marine flatworms, mating looks more like a fight than romance, with each flatworm trying to get its penis into the genital pore or even sometimes, just into the skin of the other. When any puncture leads to pregnancy, flatworm penis-

A PENIS-FENCING FLATWORM JABS ITS PARTNER (THIS PAGE, TOP), AND LEAVES A TRAIL OF WHITE SPERM (BOTTOM). □ THE HEADS OF TAPEWORMS ARE MODIFIED FOR ATTACHMENT (OPPOSITE PAGE, RIGHT). A TAPEWORM IS A STRING OF REPRODUCTIVE SEGMENTS (OPPOSITE PAGE, LEFT).

fencing is serious business. And the first to succeed becomes the *de facto* male, the other the female. In a sense, mating *is* a fight because the worm that assumes the female role then has to expend considerable energy caring for the developing eggs. For flatworms, though, like the planarians and those in the seas, ponds, or swamps, being hermaphroditic is a very good reproductive strategy, since the odds for success are doubled.

The most common kind of relationship between one animal and another, or between an animal and a plant, is that of diner and dinner.

Vicki Pearse, John Pearse,
Mildred Buchsbaum, Ralph Buchsbaum,
Living Invertebrates

The earliest flatwormlike animals were free-living creatures that became Earth's first hunters and went on to more than one-half-billion years of success. Modern flatworms, including small turbellarians and planarians, look a lot like them. Others, like Sally Wallace's record-breaking tapeworm, have taken a different evolutionary trajectory and adapted their basic body plan and behavior over millions of years to become parasites. Their survival depends upon a lifestyle that helps them find and exploit a host, but their goals remain the same as those of every other member of the phylum: Stay alive. Eat. Make more animals like themselves.

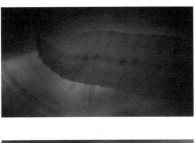

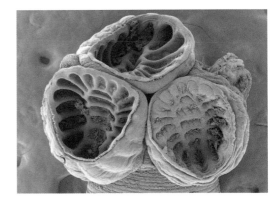

Sally's tapeworm, and others like it, take their nourishment at a considerable cost to their hosts (remember the diet-pill worms) and also depend on the conditions in the hosts for reproduction. And tapeworms aren't the only flatworms that have adapted the same basic body plan to ride along with other animals as parasites. Liver flukes, trematodes, and blood flukes, like the schistosomes that cause bilharziasis and kill 20,000 humans a year, are all variations on the same design. All have developed complex strategies for getting into their hosts, feeding, and reproducing, sometimes moving between different kinds of animals at different stages of their development, from

egg to adult, often in an endless circle. One of these, the broadfish tapeworm, typically lives in a large mammal like a bear, and leaves that host as a fertilized egg in the bear's feces, settles into a stream bed or shore sand and protects itself for the time being by hardening its exterior into a kind of a shell. Then it's eaten by a minute crustacean called a copepod, which is eaten in turn by a salmon or other carnivorous fish, which is then eaten by another bear. And on, and on, and on. Through each stage, the tapeworm egg, larva and finally the adult are well suited for survival in the intermediate hosts.

Genes survive down the ages only if they are good at building bodies that are good at living and reproducing in the particular way of life chosen by the species.

Richard Dawkins,
River Out of Eden, A Darwinian View of Life

All of our inquiry into the lives of ancient worms is really an epic investigation into our own origins and body plan, a lens into deep time that is now showing us a picture of genetic evolution. Until recently, information about the connections of ourselves and other animals to the first hunters of the ancient past came from paleontologists like Whitey Hagadorn and his trackways in the rocks, and from analyzing the relationships of anatomy and embryology among modern animals. Halfway through the last century, though, we got our first real glimpse of life's grandest engine of inheritance, deoxyribonucleic acid—DNA. Since then, a patient investigation into these instructions for life, called the genetic code, has occupied the careers of thousands of scientists, including geneticist Matt Scott.

While Whitey Hagadorn has been deciphering the external trails and traces of ancient life, geneticists like Matt Scott have been sorting through the mysteries of DNA, trying to fathom the process by which animals create their own bodies.

"All the cells in our body, with a few exceptions, have a complete set of genes. And in those genes, in that fantastically complex and mysterious genome, are the instructions for building an organism," he says, reveling in the magnitude of his quest.

MATT SCOTT (THIS PAGE, TOP). BANDS SHOW REGULATORY GENE ACTIVITY ESTABLISHING SEGMENTATION IN A FRUIT FLY EMBRYO (BOTTOM). □ HOX GENES DIRECT BODY FORMATION IN ALL BILATERAL ANIMALS (OPPOSITE).

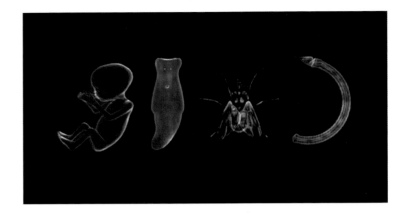

The genetic code within every living cell of an animal contains not only the instructions for building that animal, but within those instructions there is a record of the similarities shared by that animal with all of its ancestors. Think of that. If there was a first hunter with a bilateral body plan, it must have carried a set of genetic rules for making that kind of body. And if all bilateral animals are descendants of that animal, all those kinds of animals must share those same rules. In other words, all bilateral animals must have inherited that same set of rules for making the same kind of body. The genetic code that spelled 'bilateral animal' over one-half-billion years ago can still be read today in all living animals that have heads, tails and two sides that mirror each other. By comparing the genetic code of living animals, we know now that we have a lot more in common with an animal like Sally Wallace's tapeworm than is obvious at first glance.

To sort out the rules for making a bilateral body, Matt Scott looks at how the body of an animal is made when it grows from an egg to an adult. He has learned that the cells in a developing animal are told when, where and how to grow by obeying signals from special regulatory genes within the DNA. These genes trigger the activity of other genes, controlling when and where they, in turn, are allowed to do their job of growing an eye, or any other particular part of the body. In this way, a special group of these regulators, called Hox genes, pattern the head to tail axis of a bilateral body. It's like laying out a grid work to be filled in with other instructions to eventually create an entire body plan.

Regulation of the development of an animal is a bit like a set of signals that can switch themselves on or off to tell the individual genes in the massive sequence of instructions when to go into action. "Understanding how these genes build those patterns is a dream come true for a lot of biologists," says Matt Scott. In effect, to control the regulatory process of building an animal body, a small set of Hox genes is conducting an orchestra of thousands of other genes.

Making the dream come true has been a long and painstaking process of figuring out how genes create bodies by isolating the individual gene codes and looking at mistakes in the developmental process, called mutations, to figure out how they happened. In their labs, pioneer geneticists worked with tiny fruit flies because they are easy to raise by the millions and they reproduce fast to create generation upon generation of flies within a few weeks. By tinkering with their genetic codes, the geneticists created flies with legs growing out of their heads instead of antennae, or antennae growing where legs would usually be. They found out that in such mutant flies, it is the Hox gene that contains the mistake. The mutant gene produced its signal to produce a body part but in the wrong place.

This is not just a Mel Brooks Frankenstein movie with a fruit fly playing the role of the monster. By creating such mutations, the geneticists learned what the proper sequence should be to produce a leg in a leg's place, and which genes regulate events to produce that conclusion. It's just like breaking a code by trial and error. And by finding and comparing the Hox genes in lots of different animals, geneticists have found that those genes are very similar in all of them. Some Hox and other regulatory genes can actually be planted in totally different kinds of animals and still produce control signals for the very different kinds of body parts. A gene that regulates the formation of an eye in a mouse, when transplanted into a fruit fly, will trigger the proper fruit fly sequence to create a proper fruit fly eye in the proper place. Only the genes native to the particular kind of animal, a fruit fly-eye gene, can produce the end result, a fruit fly eye, even though a mouse gene cracked the genetic whip.

This was an incredible discovery. It means that a fruit fly and a mouse—an insect and a mammal—have regulatory genes that are so similar they can be interchanged. And even more incredibly, when geneticists looked at Hox genes from all kinds of other bilateral animals, they

found that their codes were amazingly similar. The Hox genes in different animals did not produce the same body plan in all of them, but they laid out a similar pattern in all bilateral animals. The startling conclusion: Even though insects and mammals have very different body plans, they must share a common ancestor whose DNA contained the first Hox genes.

That ancestor, therefore, must have been the first bilateral animal, which gets us back to Whitey Hagadorn's version of time travel. He and other paleontologists have fixed the time that the earliest tracks of bilateral animals appear on Earth, left there by a primitive version of a modern flatworm. By combining those discoveries with the story of the Hox genes' voyage through time, we know that those ancient flatwormlike creatures pioneered the basic body plan, which has been inherited by people and all other animals with heads, tails and places to go.

Before we give Tom Hills the news that it's his own relatives tearing up his farm in Scotland, we'd probably better take him down to the pub for an ale or two.

MANY MODERN FLATWORMS HAVE COLORFUL BODIES WITH STRIKING DESIGNS, BUT IT'S NOT CLEAR WHY.

ANNELIDS

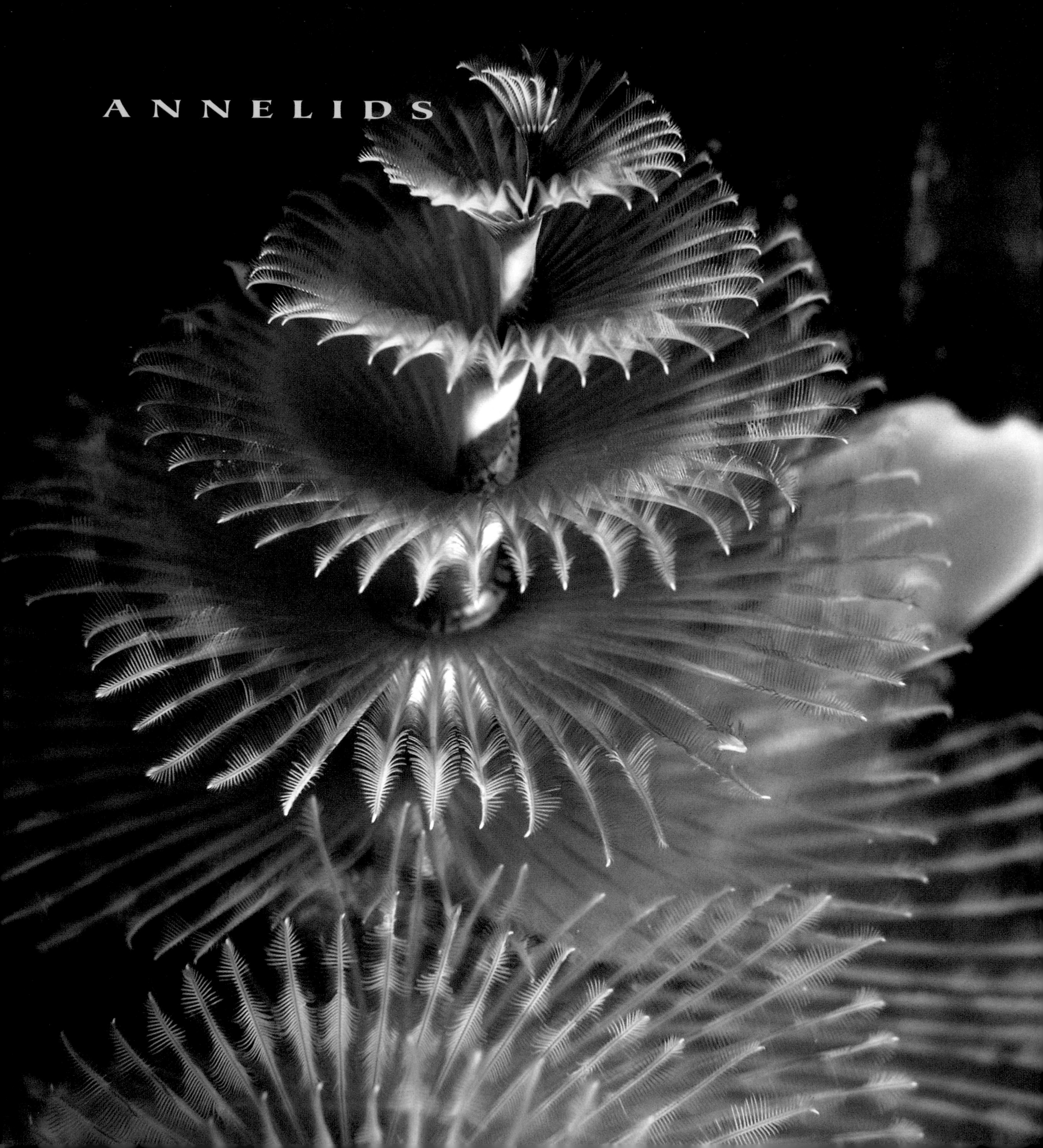

CHAPTER 4

THE NEXT SEGMENT

An Explosion of Life

What I like doing the most is actually getting out, getting down, getting dirty in the mud with the worms themselves. Getting out there whether it's pouring rain, whether it's a low tide at dawn, I don't care. I just like to be out there with the worms, seeing them in their own habitat.

Damnhait McHugh, Biologist

Imagine that you are on your hands and knees 20 feet from one of those prosaic struggles between a robin and an earthworm, like the ones celebrated in cartoons, homilies and the sketches of school children. The battle begins at dawn under an overcast sky. Hard rain threatens after a night of drizzle, a scene painted from a palette heavy with greens from the spring trees flashing their putty-colored undersides in the building wind to the expanse of lawn so dark it seems almost black beneath the morning clouds. The robin attacks, landing first a foot from the prey it spotted from above, then nonchalantly sidling to the exposed nub of flesh in the grass where, like lightning, it strikes the earthworm. The assault carries the element of surprise and quickly, two inches of worm stretches from the ground to the robin's beak. Then, as though it summons courage, the earthworm fights back, actually tugging the robin's head to the grass. And then the worm is gone. The beaten robin hops away, then flies, quickly blending into the sky. As you watch you wonder, with a curiosity as simple as a child's: How can this be happening? How can an animal as big as that bird struggle so hard to pull a couple of inches of brown goo out of the ground? What does it mean that it can't? What does it mean that it can? What questions have you forgotten to ask?

That is science. Just that. Seeking the answers to questions—many of them obvious—about what we observe in the world. Charles Darwin, who studied earthworms during his epic life, might have experienced precisely that moment with a robin and a worm, perhaps at his home near the kilns of his father-in-law, Josiah Wedgwood, whose pottery works were turning the rich clay of the Dee Valley into England's most famous plates, cups, and platters. Darwin's fascination with earthworms was fueled by the passionate curiosity we still see in scientists, and children, whether in a National Geographic special on television or at a dinner party when one of them gets loose in an entertaining, rambling monologue on black holes, fossils, genetic blueprints, or . . . worms.

Darwin lived in an era when a universe previously thought to be fashioned and fixed by a supreme being of some kind was beginning to reveal itself as governed by a set of natural laws that are not written but discovered. In the 200 years before Darwin was alive, Kepler, Galileo, Copernicus, Newton and others had already removed Earth from the center of the universe, established some of the physical rules for the interaction of things, and set a pattern for further exploration of the natural

world that depended on an open-ended, systematic method of inquiry we now call science. Charles Darwin's exploration of the natural world led him to a life-long study of earthworms. He also had special affections for barnacles, birds, and beetles in his search for the rules governing the lives of the animals inhabiting our planet. He was a geologist, too. Geology is really the story of Earth's history revealed by rocks and time, lots and lots of time, and geology would eventually figure very prominently in Darwin's own revolutionary set of rules that govern the existence of living things.

Whether you're talking about helium and hydrogen created in a primordial explosion, carbon atoms born in stars, or worms and robins fighting it out on a lawn in England, that new universe of natural laws presumes a process that begins somewhere. Once Darwin and the other scientific revolutionaries began observing processes that operate according to a set of predictable, repeatable rules, it became impossible to begin with the conclusions of myth or religion and call it discovery. You can have a hypothesis and prove it right or wrong, but you can never start with a final answer and expect to learn anything about the world around you.

Darwin's theory that living things obey a set of natural laws through time was a new lens through which we humans viewed the past. *The Origin of Species by Means of Natural Selection or the Preservation of Favoured Races in the Struggle for Life* is a description of a process. But what about a beginning? Darwin didn't know when the parade of animals began, but he knew it began somewhere, at some time, and that it left evidence of that process through time. He also knew that animal diversity appears suddenly in the fossil record, so he thought there must be a long history that wasn't evident at the time. After Darwin handed humanity a new way to look at the history of life, the evidence seemed to gush from the earth. (The genetic story of the relationship of present life to past life would begin to emerge a hundred years after Darwin's description of evolution, but it would describe a process that operates with the same set of rules.)

> *Very suddenly, and at about the same horizon the world over, life showed up in the rocks with a bang. For most of Earth's early history, there simply was no fossil record. Only recently have we come to discover otherwise: Life is virtually as old as the planet itself, and even the most ancient sedimentary rocks have yielded fossilized remains of primitive forms of life.*
>
> Niles Eldridge,
> *Life Pulse: Episodes from the Story of the Fossil Record*

Our home planet coalesced into a sphere about four-and-a-half-billion years ago, acquired water and carbon about four billion years ago, and three-and-a-half-billion years ago, according to microscopic fossils, bacterial cells began to show up. Life.

Three billion years later this primitive living stuff was flourishing. Tiny fossil sponges and their spicules mark the beginnings of multicellular animals, and soon after we can see the shadowy impressions of more complex fans and polyps and things with no names that show that animal life was in an experimental phase. Then suddenly, around five hundred and twenty million years ago, in the relatively short span of about 10 million years (given the usual pace of geologic time), life exploded in a radiation of abundance and diversity that contained the body plans of all the animals we know today.

In the first decade of the twentieth century, we found fossils that confirm this one-of-a-kind explosion of life, and as is the case with all great moments in science, there is a story. Actually, there are several versions, but in the one told most often around the campfires of paleontologists, the stumbling horse of the wife of a fossil hunter named Charles Walcott led to the discovery of a 505-million-year-old deposit of sediment that has come to be known as the Burgess Shale. Maybe it was the luck of Helena Walcott's foul-footed mount, maybe not, but in late August 1909, in the rocks kicked off a cliff face on a ridge between Mt. Wapta and Mt. Burgess, in British Columbia, her husband found and retrieved not only some new notes in the fossil record, but a soaring crescendo that would forever change the way we listen to the past.

The Burgess Shale contains the fossils not just of a diversity of animals with hard skeletons but of a suite of rare, soft-bodied forms that died suddenly when a turbulent mudslide carried an entire community of animals to the bottom of a submarine cliff and into a basin with little available oxygen. Without oxygen, the bodies never decomposed, so over time they were covered by sediments that turned to rock, creating exquisitely detailed fossils, shadows of the ancient past in what became dark Cambrian shale. Since Walcott's discovery of the Burgess Shale fossils, other paleontologists working around the world, most notably in China, have found other fossils from the same period, which means that the Cambrian Explosion occurred all over the globe.

When Charles Walcott discovered the fossils of the Burgess Shale, he was arguably the most powerful and influential scientist in America. Secretary of the Smithsonian Institution, and a Victorian gentleman, he carried the mission of discovery like a battle flag. He was also the director of the U.S. Geological Survey, president of both the National Academy of Sciences and the American Association for the Advancement of Science, and custodian of America's forest reserves. His trip to the Burgess Shale was one of the annual family outings that remain popular among paleontologists, and he had gone to the Rockies knowing he would find trilobites and their arthropod relatives.

Walcott shipped his specimens back to the Smithsonian, set about classifying them and soon declared most of them to be related to modern groups of animals, especially arthropods. Although he was way off the mark in many of the specifics of his identifications, he did the best he could, given the science and technology of his time. The possibility that the fossils of the Burgess Shale

BURGESS SHALE ARTHROPOD FOSSILS: *MOLARIA SPINIFERA* (TOP) AND THE LACE CRAB *MARRELLA SPLENDENS*, THE MOST ABUNDANT FOSSIL (BOTTOM).

represented a sudden radiation of dozens of modern animal body plans flew in the face of the current understanding of the natural laws of evolution at that time. The work of speciation seemed until then to progress at a steady rate. Gradually. The possibility that a sudden burst of evolutionary energy could occur was not then available to logic. (Even today, scientists debate the amount of time over which the Cambrian radiation occurred.) The fossils of the Burgess Shale were weird and fascinating, but they reposed in boxes in the Smithsonian and other paleontological collections for another 60 years before their true meaning was known.

All those scientists who worked with this stuff for a hundred years, they all had it wrong. . . . It's sort of a bit like pin the tail on the donkey, except you don't know if it's a tail, if it's a donkey, or which end is which anyway.

Desmond Collins, Paleontologist

Des Collins is a senior curator at the Royal Ontario Museum and since 1975, he has devoted many field seasons to reconnoitering, excavating and collecting fossils from the Burgess Shale of British Columbia. Every summer he makes the spectacular drive through the Canadian Rockies to the town of Field, just west of the continental divide, descending along the Kicking Horse River to the narrow pocket of the Yoho Valley. The Burgess Shale quarry is now a part of a Canadian national park, but also designated by the United Nations as a World Heritage Site, a place considered so precious and significant to the history of Earth that it is tended like a forbidden shrine. Unless you are Des Collins, or part of his research team, or the occasional film crew, or one of the handful of other people allowed to make the guided trek from Field to the Burgess Shale during the two months of the year when the snow clears, you can't actually go to the quarry.

But if you could, you would hike for three hours from a trailhead that begins in Field to an elevation of 7,546 feet where Fossil Ridge runs northwest–southeast from Wapta Mountain to Mount Field. The last 300 feet or so is the real test, when you must scramble up the steep scree after a long trek, but then you're at Walcott's Quarry. If you can tear yourself away from the presence of the world's most famous fossils, the view is breathtaking. Dead ahead to the west is Mount Burgess, to the right, below, is the shimmering, milky green splash of Emerald Lake, and down to the left, you

WALCOTT'S QUARRY AT THE BURGESS SHALE, B.C., 1924 (ABOVE). □ FOSSIL *AYSHEAIA* (OPPOSITE PAGE, TOP) AND RECONSTRUCTION OF *AYSHEAIA* (BOTTOM).

can pick out the Trans-Canada Highway running along the edge of the braided course of the Kicking Horse River. The quarry itself is about the size of a boxcar, with the chips and slabs of excavation scattered on the down slope, and there, in August of 1909, strange creatures from the distant past whispered to us about an explosive beginning to a half-billion years of animal diversity.

You can hunt fossils in the field, where you also get to swat flies, eat bad food, hike long distances, and sleep on hard ground. Or you can hunt in the cool, dark storage cases in museums. Walcott and his assistants mined about 80,000 fossils from his quarry in the Rockies, and though they were a sensation and were loaned by the Smithsonian to paleontologists around the world, nobody really took a hard look at them. Finally, in the early 1970s, the meaning of Charles Walcott's fossils was revealed, not by the accident of a stumbling horse, but by the patient and systematic inquiry of a paleontologist named Harry Whittington, whose fascination with ancient invertebrates led him to the cabinets where the Burgess treasures were stored.

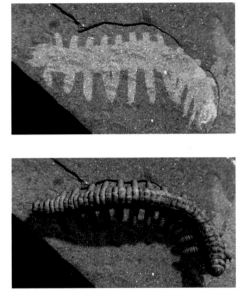

Whittington concluded that some of the creatures in the Burgess fossils were indeed arthropods, the ancestors of modern crabs, spiders, and insects, but the rest were ancient examples from the 35 body plans—or phyla—documenting the sudden appearance of an astounding diversity. The Burgess Shale also contains, Whittington said, not only extinct species of ultimately successful types of animals, but entire forms of life that departed the planet forever, even as the Cambrian was exploding. The sudden feast of diversity, it seemed, was even more sumptuous than Walcott ever thought possible.

The confusion about the animals of the Burgess Shale is quite understandable before Whittington, Des Collins and others began to decipher their true meaning. Few of the fossils reveal an entire animal, so scientists must recreate ancient creatures piece by piece.

"You just never know what you're going to find. So the excitement is finding something completely different and saying: 'Gee Whiz! Look at that! I've never seen anything like that,' " says Collins. "But perhaps even more exciting to me is finding something that I'm looking for. I know I have some pieces of a puzzle, so I know there's an animal out there that has this particular structure, or looks a certain way. I want to find the animal all put together. I remember those moments vividly, when you get something and it's never quite what you expect, but it's sort of the last piece of the puzzle."

It often takes years and several false starts to piece a single creature together from various fossils. It's a tricky process of trial-and-error, as Collins found with a big Cambrian predator named *Anomalocaris* (a-nom-uh-luh-car-is). Nothing like this animal exists today, though its basic body plan is still used by living arthropods. "*Anomalocaris* was first described over a hundred years ago in 1887, and at that time, it was based on a claw fossil," Collins recalls. "This was thought to be the body of a shrimp. It seemed to lack a head, but it had what appeared to be legs on it." Another fossil, thought to be a separate animal, looked like the impression of an ancient jellyfish, but these two creatures

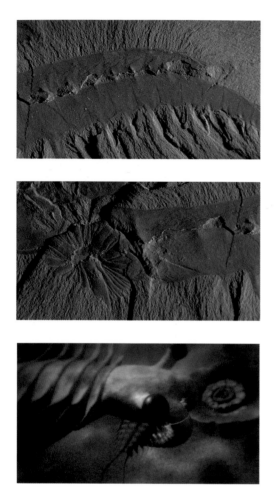

proved eventually to be part of the same beast, *Anomalocaris*. "What we thought was the body of a shrimp was actually claws, and what was thought to be a jellyfish was actually the jaws of this much stranger animal," Collins says.

"What we've since collected is a complete specimen of *Anomalocaris*. So we have a pretty good idea. We've got the tail, we've got the head, and here we have a nice model of the whole animal," Collins goes on. "This is the major predator from the main Burgess Shale site. And we even have claws, which are twice this size, so it's conceivable that *Anomalocaris* got up to three or four feet in length. So, this was a major predator compared to all of the other animals of that particular time. It took over a hundred years to work out what this animal looked like from the first piece that we had."

"And all those scientists who worked with these fossils for a hundred years, they all had it wrong," Collins says. "So, of course, that makes me very nervous that when I'm working with this stuff—particularly if I have something that seems to be a piece of something—since I cannot relate it to something that's alive today, when I try to put it together, the chances are I'm going to be wrong." The self-correcting process of science continues.

And whether or not the pieces of all the animals of the Cambrian Explosion fit together, the great message in those fossils remains. Earth is four-and-a-half billion years old, but all the basic body plans of most of the animals that ever lived appeared within a mere 10 million of those years. (The root animal groups of sponges, cnidarians, and flatworms almost certainly appeared before the Cambrian.) This is almost too much to be believed, but subsequent study of the Burgess fossils and others from sites in the Cambrian horizon confirm that this period was so rich in diversity that we can call it an

explosion of life, a beginning for lineages that still exist over one-half billion years later.

Each of those lineages is called a *phylum*, from a Greek word meaning "race." Because scientific inquiry is constant and self-correcting, debate continues about just how many phyla exist, but 35 is a pretty good number. Most of the 35 are pretty small clubs, and about 90 percent of all the billions of species of animals

that ever lived can be grouped into just eight of them. Those eight—sponges, cnidarians, flatworms, annelids, arthropods, molluscs, echinoderms, and chordates—are the stars of this book because they represent virtually the entire animal kingdom.

All members of each phylum exhibit anatomical traits similar enough to show a kinship of body architecture that binds the group together. Earthworms, like that tough little warrior on the lawn, for instance, belong to the phylum Annelida (their name comes from a Latin word meaning "little ring"). Annelids are distinguished by ringlike external bands that coincide with internal partitions dividing the body into segments, each containing a repetition of kidneylike organs, nerves, muscles, and blood vessels. With circulatory systems to distribute blood and oxygen and one-way guts, their bodies are enormously more complex than modern flatworms. A gut that goes from one end of the body to the other is among the cleverest of inventions in the evolution of animals. With such a gut, food can be continuously taken in by a mouth, processed as it passes through the body and released as waste at the other end. Not only could the first annelids continually digest their food, but they could squirm, crawl and slither as they did it, thus not interrupting their movement, whether hunting or fleeing. And annelids are active predators and scavengers, many burrowing incessantly as they make their living in water and on land. Modern members of the annelid phylum include earthworms, marine and terrestrial bristle worms, and leeches, about 15,000 species in all.

One of the early examples of the annelid body plan, a worm called *Canadia spinosa*, appears in the fossils of the Burgess Shale. Although it is not necessarily the direct ancestor of any modern worm, it is a representative of that race of animals that has endured through millennia.

> *One of the amazing things about the animal kingdom right from the Cambrian on is that there's only about 35 body plans, basic designs, yet there are millions of species representing everything from insects to whales. Animals that live in water, animals that burrow under ground, animals that live in coral reefs, animals that swim in the ocean, animals that live on the Antarctic ice.*
>
> Rudolf A. Raff,
> *The Shape of Life: Genes, Development, and
> the Evolution of the Animal Form*

It was as if nature struck upon life's essential designs in a single evolutionary leap. Every new shape of life that has followed the Cambrian Explosion has been a variation on those basic

FOSSIL *ANOMALOCARIS* CLAWS (OPPOSITE PAGE, TOP) AND JAWS (MIDDLE) WITH RECONSTRUCTION (BOTTOM). DES COLLINS (LEFT). □ FOSSIL ANNELID *CANADIA SPINOSA* (ABOVE).

architectural themes, and perhaps more remarkably, no new body plans have evolved since. Rudolf Raff, a professor of biology at Indiana University (and the author of a book whose title may sound similar to this one) is among the legions of men and women who have devoted their lives to sorting through the effects of so marvelous a beginning. Like many biologists, he was drawn into the vortex of science as a child, and his curiosity continues to fuel what is now groundbreaking research on the relationships of animals through time. "What is a body plan? Well," he says, "a body plan is a concept that we made up. And what I mean by this is not that they don't exist, but rather that we recognize that different kinds of animals resemble each other. When we divide them up this way, we realize that there's actually sort of an underlying plan to groups of animals. There are themes of construction."

"Imagine a machine like a Model T Ford," Raff continues. "It's a pretty simple car. The cars of today are the evolutionary descendants of the Model T Ford, more elaborate, but you recognize the body design in there. So it is with the evolutionary animal kingdom. Evolution has produced millions of species of animals. There are probably somewhere between 20 and 30 million kinds of living species right now, and the same was probably true in the past at any slice of time that you take. So in the 500 million or so years since the Cambrian, there has been an immense wealth of animals, everything from dinosaurs to bats to earthworms, and it's all happened within these body plans, 35 or so body plans and millions and millions of species."

First, a question of origins: How could so much anatomical variety evolve so quickly? In particular, must novel evolutionary mechanisms be proposed for such a burst of activity?

Stephen Jay Gould,
"Showdown on the Burgess Shale,"
Natural History Magazine

How *did* all the basic themes in animal architecture appear within just 10 million years?

"We really don't know," says Rudy Raff, in that manner of scientists, and children, for whom questions can be much more interesting than answers. But we do have some good clues. "It would seem likely that there's not a single cause for a unique event like that, but there have been a number of hypotheses put forward. Good ones."

First, the bloom of diversity might have been a genetic revolution in which the genes that regulate the basic patterning of a body plan had to arise before an animal could make a head, a brain, or features like legs and arms and claws that had to be correctly organized to produce a complex body. Suddenly, in the Cambrian, the genetic code reached sufficient control over the production of not just simple animals, such as sponges and jellies, but the more complex races of creatures.

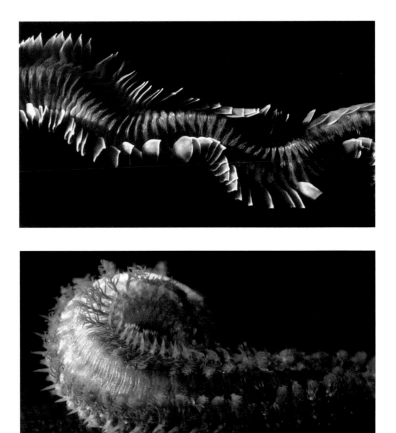

A second explanation for the Cambrian Explosion may lie in a radical shift in the nature of the planet itself, most likely in the oceans, which were home to all living things at the time. The most obvious of these ecological transformations is a change in the oxygen levels, or a reduction in the levels of other gases that had until that time inhibited growth and diversity. And finally, the Cambrian Explosion might have happened because the food web itself became complex enough to provide a multitude of possible ways to make a living and out of this, an arms race for control of that food developed. As more and more animals evolved, that arms race intensified in the battles of the hunter and the hunted.

Perhaps it was the coincidence of ancestral worms evolving mobility and the power to hunt their prey that stimulated other forms to defend themselves, which created

ANNELID'S HEADS OFTEN HAVE SENSORY TENTACLES AND EYES (OPPOSITE PAGE). □ ANNELIDS HAVE BODIES WITH REPEATING SEGMENTS AND APPENDAGES (RIGHT).

forces for evolving stronger predatory skills and equipment. Because there is an advantage in eating and avoiding being eaten, the arms race would never end. Perhaps, too, nerves that enable sensing and the coordination of motion, pioneered most primitively by jellies as networks and in the more sophisticated central systems of worms, allowed the building of bigger, faster bodies that fed the arms race. Size is a great advantage for a predator. Once a big, fast-swimming, deadly character like *Anomalocaris* was on the scene, the drama of life took on a potential for sudden death that had been absent until that time.

The greatest architectural breakthroughs of all the Cambrian animals were substantial bodies against which muscles could work to produce powerful movements. The secret of the success of these new bodies was an internal cavity that not only provided room for large complex organ systems but proved essential for more complex ways of living. The annelid body is the simplest expression of this new body design. A worm is basically a hollow tube with a fluid-filled internal space to create a hydrostatic skeleton. We have bones; an annelid's skeleton is made of fluid. These hydrostatic skeletons were pioneered by anemones, which filled their stomachs with sea water to support their bodies, but an annelid could carry its hydrostatic skeleton with it all the time. While the earliest annelids lived in the ocean, this kind of permanent hydrostatic skeleton would be essential when the worms took up life on land. An elongated worm shape would also prove to be one of the most successful body designs. This shape is so common on our planet that if intelligent aliens capable of assuming any body plan were to land on Earth, they would most likely transform themselves into worms to blend in.

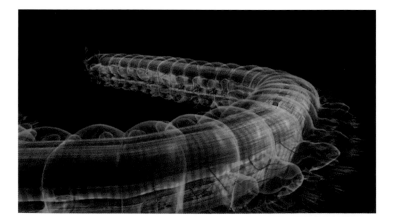

Aliens might also have transformed themselves into worms simply because they are beautiful. Biologist Damnhait McHugh loves worms the way others love birds, butterflies or tropical fishes. "I think it's very easy to get people excited about worms if they can appreciate the diversity, the color, the different lifestyles, the different ways that worms feed," McHugh says. Anyone willing to take up worm-watching as a hobby would be dazzled by what they see.

Modern annelids live in every habitable niche on Earth except the sky, from the most mundane patches of suburban lawn to hot, deep ocean vents. Of all the annelids, the marine polychaetes are the most uniquely and strikingly beautiful. They come in a startling spectrum of colors, and if you look closely you can see the intricacies of their durable bodies, the beautiful, constantly moving bristles that help many of them get around, and the feathery arrays on the heads of some that would be the envy of any costume designer.

But whether you think worms are beautiful or not, form follows function in nature and the basic mechanical equipment for making a living in the ground is an elongated body with a hydrostatic skeleton. A worm. But another annelid feature—segmentation—would make annelids not only the ultimate burrowers but also highly efficient at moving across the ground with a wiggling motion. Annelids consist of externally and internally repeating segments, except for the head end, which contains a mouth and sense receptors for establishing direction, and the tail end, which contains an opening through which waste passes. The hydrostatic skeleton is thus divided into separate compartments. Earthworms, like many other annelids, creep along or burrow by coordinating two sets of muscles, one longitudinal and the other circular. These muscles work against the non-compressible fluid of the hydrostatic skeleton, altering the shape of these separate compartments. It's like squeezing a chain of water balloons. When the circular muscles of a region contract, that part becomes thinner and elongates. Contraction of the longitudinal muscles causes the segment to shorten and thicken and the worm moves forward as alternating contractions of circular and longitudinal muscles progress along the segments like waves.

In an earthworm, for instance, each segment has four pairs of setae, or bristles, that provide traction as the waves pass over the body. When the robin strikes, the worm digs in with its bristles and ripples its muscles, segment by segment, to back deeper into its burrow. Obviously, burrowing into the mud to flee, a strategy still employed by earthworms against robins, has always been a great tactic for staying alive. The appearance of burrowing worms, the annelid ancestors of earthworms that can defeat the efforts of much larger animals bent on eating them, was a enormous step in the process of animal diversification.

Burrowing worms thrived in ancient seas not only because they could safely get away from a predator, but because burrowing brought the worms to rich new feeding grounds. Dead organisms and organic debris continually settle to the seafloor, but it was not until worms could burrow into these sediments that animals could eat this enormous treasure of food. Annelid predecessors were sponges, jellies, and simple worms, none of which have a gut for passing food from a mouth to an anus. The annelid's one-way gut becomes a tube within a tube, a highly effective arrangement because the worm can keep burrowing as it processes its food through its gut. If you eat by digging through your food, this is awfully important.

DAMNHAIT MCHUGH (OPPOSITE PAGE, TOP). ANNELID BODY PLAN SHOWING LONGITUDINAL MUSCLES AND REPEATING INTERNAL ORGANS (BOTTOM). ☐ FREE-LIVING POLYCHAETE (THIS PAGE, TOP), EARTHWORMS (MIDDLE), AND TUBE WORM (BOTTOM).

As an earthworm burrows it eats, getting nourishment from bacteria and debris in the soil through which it is passing. While the effect of a single worm seems negligible, the cumulative effect on Earth is staggering. A feeding worm takes in dirt and plant debris, extracts food, and releases castings into the surface, effectively performing as a recycling engine for nature's most vital element, carbon, as it releases carbon dioxide. Collectively, billions and billions of industrious worms have contributed to keeping Earth healthy. If all the material that has ever moved through earthworms alone was piled up on the surface of the globe, the heap would rise thirty miles, more than five times the height of Mt. Everest.

Without worms, Earth might be a very different, and less hospitable, place. According to Paul Hoffman, a professor of geology at Harvard University, from one billion to 750 million years ago there was only one supercontinent, Rodinia, surrounded by a vast ocean covering 70 percent of the globe. Then Rodinia began to break up into smaller continents, creating many more miles of shoreline. One result was that more of the carbon in living organisms was buried in the sea. Volcanoes, the main source of carbon, could not provide enough carbon dioxide to maintain the level in the atmosphere needed to keep Earth warm.

As a result, the polar ice caps expanded, covering a large part of the surface with glaciers. At that point, global cooling accelerated because ice, being white, reflects heat back into space much more efficiently than darker colors. If more snow reflects more sunlight back into space, this soon causes runaway glaciation. The ocean freezes over. There is mass mortality, but there are survivors.

The ice age cannot last, however. Volcanoes keep belching out carbon dioxide, which the frozen ocean can't absorb. Carbon dioxide gradually builds up in the atmosphere until Earth begins to warm again. The ice melts, releasing water vapor, another potent greenhouse gas. Open, less reflective water returns. "Then all hell breaks loose," Hoffman said. "The melt back is extraordinarily rapid. It could be less than 1,000 years. Now the survivors have to face the heat."

Such a catastrophic event, producing a radically different environment, could have played a role in the Cambrian Explosion of diversity. Asked if Earth could turn into a snowball again, Hoffman said no. There are now so many organisms like annelids recycling carbon into carbon dioxide that the supply cannot drop low enough to allow the ocean to freeze.

Earth is protected, it turns out, by worms.

We have to keep in mind that even creatures like the lowly worm, we depend on them. We depend on their diversity. We depend on the role they play in the ecosystems. We should see ourselves, I think, as custodians of the great diversity of animals that we see around us today. But we should also remember that we are not masters of this diversity. In fact, we depend very much on the diversity of animals around us . . . even the lowly worm.

Damnhait McHugh, Biologist

TUBE-DWELLING ANNELIDS USE TENTACLES AND FANS TO EXTRACT FOOD FROM THE WATER.

ARTHROPODS

C O M I N G A S H O R E

The Conquerors

Life's transition from the sea to the land was perhaps as much of an evolutionary challenge as genesis. What forms of life were able to make such a drastic change in lifestyle?

Jane Gray and William Shear,
"Early Life On Land," *American Scientist*

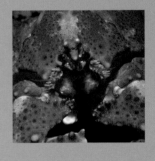

All life on Earth began in the sea. When the first animals made landfall, obviously they had to evolve ways to move around and breathe in a place where oxygen is delivered in a much different way than in water. When we try to go back to the sea, our chief concerns for survival are figuring out ways to breathe, maintain body temperature, move around, and sometimes, avoid being eaten. The human desire to submerge in water and still be able to breathe has probably existed as long as we've been diving to gather food, repair (or sink) ships, and marvel at the wonders of a world so different from our own. Getting past the fear of the unknown that lurks in the water is no small feat, since the sea is as alien to us land-dwellers as the surface of Mars, and is full of monsters that are, of course, waiting to kill and eat us.

Still, early on we hungered for a look into that utterly hostile environment, and as soon as our ancestors overcame the terror of the deep, we discovered that we could submerge for a frantic minute or two by simply holding our breath. Breathing through reeds and tubes is an obvious second step, but that limits us to a few feet because the pressure differential makes it impossible to suck air much beyond that depth. (Try it sometime.) Breathing from bags of air doesn't work, either, because the build-up of carbon dioxide will kill you pretty quickly. By the sixteenth century, divers were using suits of leather and surface pumps that forced air through hoses to remain at depths of up to 60 feet for a few minutes at a time. That technology evolved in the twentieth century when hard-hat divers routinely ventured beneath the sea to work and explore. Submarines and diving bells followed, some served by air pumped from the surface, some by compressed air carried in tanks on the deep-sea craft themselves.

Finally, in the mid-1930s, an American aviator named Guy Gilpatric living in southern France pioneered the use of rubber goggles with glass lenses for skin diving. Swim fins were invented by Frenchman Louis de Corlieu. In 1938, Gilpatric wrote about his new diving gear and exploits in *The Compleat Goggler*, the first book on underwater diving and hunting. Among the book's readers was a young French naval lieutenant named Jacques Cousteau, who in collaboration with engineer Emile Gagnan would eventually invent and patent the Aqua-Lung, a demand-regulator to be used with tanks of compressed air. He became the world's first enviro-celebrity, and ever since, humans willing to wear rubber suits, masks, fins, buoyancy control devices, and tanks have been breathing, hunting and exploring under water. . .but that's the best we can do.

> *Adapting to life on land is still a more difficult step. Temperature ranges are even more extreme, drying is a constant threat and respiratory mechanisms must be modified to use air.*
> Ralph and Mildred Buchsbaum and John and Vicki Pearse,
> *Animals Without Backbones*

The history of the animals whose descendents would be the first to breathe air and live on land begins over one-half billion years ago, when the earliest members of a race called arthropods inherited their biological architecture from more primitive, bilateral animals. The earliest arthropods took the basic ingredients of a segmented body—appendages, tissue, organs, nervous system and brain—and diverged to become entirely new kinds of animals.

These creatures quickly became powerful hunters and scavengers, and established patterns of adaptability and dominance that continue into the present.

One group, the trilobites, were predators in the sea for more than 300 million years, beginning in the Cambrian when their legs, complex armored bodies, and the first true eyes in the animal kingdom gave them immense power over less-endowed prey. Trilobites look a bit like modern pill bugs common in gardens, but they are only distantly related. Still, just about everybody knows what a real trilobite looks like because there were simply so many of them and because their hard shells endured through time as perfectly preserved fossils. Trilobites have been mined from quarries by the millions to be sold as paperweights, pocket rocks, tie tacks, cuff links, necklaces, and bracelets in just about every rock shop in the world. They flourished in the

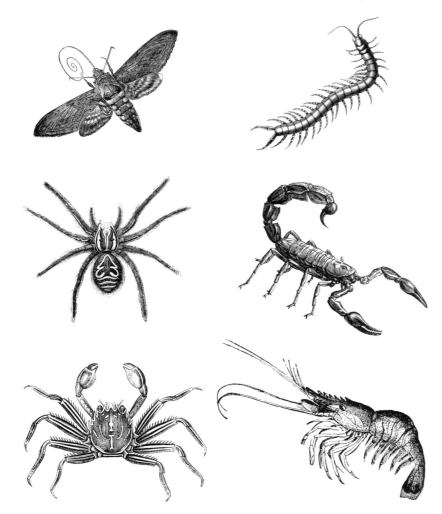

VARIOUS ARTHROPOD GROUPS (THIS PAGE): INSECT AND CENTIPEDE (TOP TO BOTTOM, LEFT TO RIGHT), SPIDER AND SCORPION, CRAB AND SHRIMP, AND TRILOBITE (ABOVE). □ MARINE ARTHROPODS (OPPOSITE PAGE, SMALL PHOTOS, TOP TO BOTTOM): CORAL CRAB, SLIPPER LOBSTER, HERMIT CRAB, AND SPIDER CRAB AND (RIGHT) KELP CRAB.

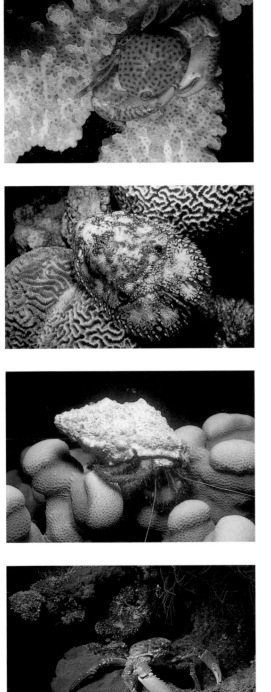

Paleozoic and achieved immense diversity, branching into thousands of species over time.

Arthropods were and continue to be nature's greatest generalists, which has contributed immensely to their success both in the water and on land. The more specialized a group of animals is, the less able it is to cope with change once the inevitable happens and old habitats change beyond the point of supporting a particular life and body style.

> *Whatever criteria for success one cares to adopt, the animals with legs keep coming out at the top of the list. Whether one rates success in terms of numbers, number of species, range of habitats exploited or simply as total mass of animal tissue, the limbed creatures clearly have the advantage.*
>
> Martin Wells,
> *Lower Animals*

The arthropod body plan has adjusted itself many times since it first appeared, adapting to the pressures of changing environments and the dance of predators and prey, but then, as now, it is defined by a segmented body, jointed legs, and a hard exoskeleton. The name *Arthropod* means "jointed leg." From its Greek root we also get the modern word *art*, which in its conception was defined as a joint between reality and abstraction.

The segmented body is a powerful arthropod evolutionary tool that is common elsewhere in the animal kingdom. People are segmented, too, though not as obviously as earthworms or lobsters, and you have to know how to look at our spine and muscles to see the segmentation. The arthropod kind of construction allows a body to build itself in identical segmented units that group together for functions in a process called tagmosis.

Many of the tasks of living, like eating or sensing the world in the water or on land, are handled by appendages attached to each segment of the body or the head of an arthropod. These appendages, in turn, are jointed, which makes them both flexible and easily altered to become antennae, claws, jaws and other mouth parts, and especially legs—legs for walking, legs for clinging, legs for holding onto egg sacs, legs for swimming, legs for digging. Sections of the legs even evolved into gills for obtaining oxygen from the surrounding sea water. These corresponding appendages are said to be serially homologous, that is, similar in derivation on each segment but not necessarily similar in function. Of course, legs were excellent for making a living in the sea, but along with

THE ARTHROPOD JOINTED APPENDAGE HAS EVOLVED INTO A FANTASTIC ARRAY OF TOOLS AS SEEN IN THESE MARINE CRUSTACEANS.

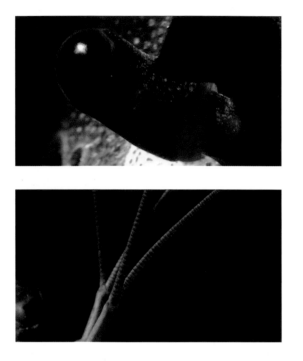

other arthropod adaptations, they worked fine on land, too, allowing arthropods to counter the force of gravity and hold their bodies off the ground. And the joints would prove to be excellent shock absorbers for walking over rough terrain. Instead of just slithering and crawling, legs allowed arthropods to jump and walk and run at high speed and, eventually, even evolve wings to fly.

> *There are good biological reasons why insects aren't the size of mobile homes and Greyhound buses and the like. One thing that keeps them small is their exoskeleton. It has to be molted and this presents insurmountable difficulties to insects as they get bigger.*
>
> May R. Berenbaum, Entomologist

Arthropods were among the first bilateral animals with exoskeletons, in their case shells made of a nitrogen-rich sugar called chitin, reinforced with protein. In the simplest of terms, this suit of armor keeps the inside in and the outside out, and it also anchors the muscles of the claws, legs and antennae so they can do work. Inside, an arthropod has all the organs of a complex higher animal, but its blood just sloshes through loose channels surrounding these organs, unlike our closed circulatory system that distributes blood throughout our bodies under pressure. Outside of an arthropod is, well, the rest of the world, so the hard shell offers protection from predators and environmental conditions that might harm the animal. An arthropod controls the intake of water just like we do, and while it is moist on the inside, the sea does not flow through it freely. Crustaceans like crabs and lobsters go a step further in reinforcing their armor with calcium to make it extra hard. Insects will ultimately waterproof their exoskeletons.

One of the enormous drawbacks to life with an exoskeleton is that once it forms and hardens it doesn't grow, unlike our own internal bones, which are surrounded by flesh and organs and grow as our bodies grow. It's impossible, therefore, for the animal to grow for very long without just filling up the shell. When their insides grow too big for their outsides, arthropods deal with this apparent design flaw by shedding their exoskeletons in a process called molting (a.k.a. *ecdysis*, which means "getting out"). It's one of the most fascinating tricks in the animal kingdom. When a crab, for instance, is

THE ARTHROPOD COMPOUND EYE (THIS PAGE, TOP). LOBSTER ANTENNAE (MIDDLE). LEG FLEXING AT JOINT (BOTTOM). □ A BLUE CRAB PULLS ITSELF OUT OF ITS OLD EXOSKELETON IN A PROCESS CALLED MOLTING (OPPOSITE PAGE).

plumped up against its shell, the secretion of hormones controlled by special molting glands triggers the separation of the old exoskeleton from the skin and the formation of a new exoskeleton beneath the old one. The crab's old exoskeleton splits open across the back edge and loosens from every part of its body, legs, and claws. The crab then slips out of its old shell in about 30 minutes. Some take more or less time to do the job, but before the new exoskeletons harden, they quickly expand their bodies by imbibing water to make sure the new hardened skeletons have extra room for growth. While all this is underway, arthropods are extremely vulnerable and very likely to become a meal, as anyone who loves soft-shell blue crabs can tell you.

Even though arthropods have solved the problem of growing from juvenile to adult by periodically molting, the limitations of their exoskeletons are one reason Hollywood nightmare visions of giant insects terrorizing the world could never be real. The fossil record tells the story that when land arthropods were first competing and size would have been an advantage, they never got larger than a few inches in body diameter. The arthropod way of growth by molting and how they supply oxygen to their bodies probably sets their size limit.

Among the smallest of arthropods is a mite that lives on your eyelashes and the largest is a crab with a twelve-foot leg span. All the larger arthropods still live in the ocean, where the buoyancy of water reduces the force of gravity. In the ocean, too, gravity does not deform the new shell of a molting arthropod before it can harden into the right shape. But what limits the size of the most successful arthropods on land, the insects, is the way they get their oxygen. Insects don't use an internal high-pressure system of circulating blood like we do to distribute oxygen to all parts of the body. Instead they depend on oxygen gas diffusing through a system of tubules. This is a comparatively slow process if it has to go a long way and could never supply oxygen fast enough to sustain a large body.

Take the simple matter of breathing. . . . Gills are simply useless for taking oxygen from the air.

Niles Eldridge,
Life Pulse: Episodes from the Story of the Fossil Record

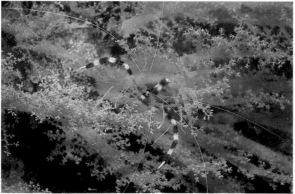

Even when the arthropods were confined to the oceans, the evolution of several breathing strategies was clearly underway. In water, which contains 40 times less oxygen than an equal volume of air, the standard tools are gills. These vary from very small, waving tentacles in small marine animals to the much larger, more extravagant arrays of feathery structures in larger creatures such as lobsters and crabs (and, of course, fishes). A simple oxygen concentration gradient between the interior of a respiratory system and the open water allows oxygen molecules from the water to diffuse through a thin membrane. The oxygen dissolves in the blood fluids and is then whisked away to all parts of the body. Some marine arthropods, including the scorpion ancestral line, breathe with a system called book gills. These are much less complicated than other gills, really just a row of flaps that are kept in constant motion so an ample supply of dissolved oxygen can enter the respiratory system. During their time as exclusively marine animals, along with all the rest of life on the planet, the arthropods became generalists when it came to breathing. The evolution of several breathing strategies was clearly underway.

So with segmented bodies that turbocharge their ability to accommodate environmental changes, with legs that let them walk, with hard shells that keep water in or out, and with highly adaptable respiratory systems, arthropods were prime candidates for colonizing land. It was only a matter of time.

Some of the early signs of animal life on land look like scratches in rocks that had formed a little over 400 million years ago during the early Devonian, a time named after the part of the British Isles where those rocks were first discovered and dated. These scratches are very different from the trackways left by worms on the seafloor because they were clearly made by animals with legs, most likely relatives of the scorpions called eurypterids. By the time they came ashore, whether fleeing enemies, in search of an easy meal, or to find a safe place to lay their eggs, the eurypterids had already established themselves as fierce and successful killing machines in the ocean. Some of them grew as long as six feet, equipped with legs, swimming paddles and fast claws armed with spikes. As Simon Braddy, a paleontologist who studies these ancient monsters put it, "I'd much rather be in a pool with a six-foot shark than a six-foot eurypterid."

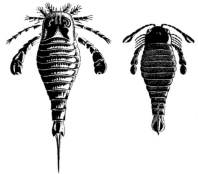

There is little doubt that eurypterids made landfall even before the Devonian. They were well equipped to pioneer life on dry land with the basic arthropod tool kit: legs, a hard shell to keep their bodies from drying up, and book gills that could function while merely moist and not completely immersed in water. But these adventurous eurypterids never completely abandoned their home in the sea to become terrestrial. Their relatives, the scorpions, however, would eventually become fierce predators on land.

The first real colonization of land by animals that stayed would be made by much smaller explorers. A series of relatively recent fossil finds reveal a complicated sequence of events. There are four major forms of early terrestrial arthropods—insects, millipedes and centipedes, spiders, and scorpions—so we know that their great race of animals had diversified before they made landfall and pursued parallel evolutionary trajectories both in the sea and ashore. And now there is ample evidence that this arthropod race invaded the land not once, not twice, but many times over.

HERMIT CRAB AND BANDED CORAL SHRIMP (OPPOSITE, TOP TO BOTTOM) ☐ EURYPTERIDS (ABOVE).

*What forms of life were able to make
such a drastic change in lifestyle?*
Bill Shear, Biologist

Bill Shear is a biologist whose interest in the evolution of insects and spiders led him into the search for their origins through paleontology. And as he examined the histories of the earliest land-dwelling arthropods, he also paid particular attention to the conditions that existed on Earth at the time they left the sea. "The world of the Devonian was probably different from our world in many, many aspects," he says. "The year was shorter and the days were not 24-hour days. You'd have a different atmospheric composition. The plant life would look different, strikingly different." It was among fossils of this first plant life that the tiny remains of those first invaders were found.

Obviously, the arthropod colonization of land was not just a matter of a couple of bugs deciding to leave the ocean for a totally alien place. Shear suspected that some sort of an ecological bridge might have been involved in making the transition, some environment that was not quite water and not quite land. And indeed, for billions of years, bacterial and algal mats had existed in the shallows along continental margins. Although this scum probably provided the perfect transitional environment, the first tiny land arthropods were found among primitive land plant fossils. To extract these minute fossils for study, Shear used hydrofluoric acid to dissolve the rock surrounding them, leaving only organic fragments containing the remains of both the plants and minute animals. "I can remember seeing some of the really striking fossils for the first time," he recalls. "I got a feeling of excitement that's probably very similar to scoring the big touchdown at the homecoming game. You just feel on top of the world, and it makes it worth all of the tedious searching and work that leads up to that." After he succeeded in extracting the microfossils from the rock, Shear painstakingly pieced together the chips and scraps of these tiny animals and found a whole suite of arthropods, including a creature not unlike a modern spider. The rock could be dated, and the addition of that clue meant that these

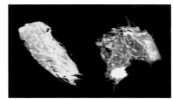

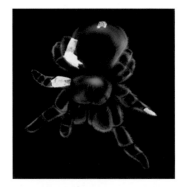

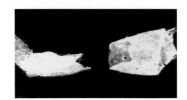

microscopic arthropods had made landfall a few million years after the eurypterids' first brief excursion onto dry land. In separate but similar forays, the arthropods had launched themselves into the new world.

And then what happened?

> *Since the invention of flight by insects probably allowed for their enormous diversification, I would say that it was one of the most important events in the history of the Earth.*
>
> Bill Shear, Paleontologist

Arthropods took over Earth.

First, with their highly adaptable body plans, they evolved a way to breathe air. The scorpions modified their book gills to become book lungs, very similar biological devices that could evolve slowly in the damp coastal zone. When the insects showed up, their bodies solved the problem of breathing air with a tracheal system that essentially allows them to mainline oxygen directly into their muscles without passing it through any gill-like or lunglike structures. Air enters their bodies through minute pores on their undersides, near the junction of their appendages and their shells, and it diffuses through their bodies in a kind of open branching system of fine tubules. An ample supply of oxygen as a gas reaches every cell of the insect's body. This kind of internal plumbing makes insects and other arthropods intensely aerobic, feeding their muscles enormous doses of oxygen and allowing them to move very quickly. Some spiders and millipedes would evolve tracheal systems, too, an example of the dynamic parallel evolution of which arthropods are so capable.

Once the arthropods were air breathers, their dependence on damp coastal environments ended, and they began to move inland. They fed on the rich detritus of organic plant debris, much in the same way that burrowing worms thrived on the debris sediments in the sea. Living plants provided limited potential for nourishment because their cells have hard walls made of cellulose that was difficult for the first waves of tiny bugs to penetrate and digest. But the invading arthropods found plenty of debris already broken down by fungi and bacteria that was still nutritious. Naturally, with their ancient histories and skills as carnivorous predators, many of the first land-dwellers ate each other, dead or

BILL SHEAR (OPPOSITE PAGE, TOP). FOSSIL FRAGMENTS ARE PIECED TOGETHER TO REVEAL AN EARLY SPIDER (OPPOSITE PAGE, TOP, MIDDLE AND BOTTOM). □ A DAMSELFLY EMERGES FROM ITS NYMPH STAGE AFTER A SIMPLE METAMORPHOSIS (THIS PAGE, TOP). DRAGONFLIES ARE ACROBATIC PREDATORS OF THE SKY (BOTTOM).

alive, and the same arms race that produced dramatic evolution in the ocean began on land.

Among their adaptive strategies, some arthropods evolved as hunters in fresh water, reverting to exclusively aquatic lives. Others came up with multiple developmental stages as predators in the water, on land, and in one of the great radiations in nature, in the air. Learning to fly was the *tour de force* of the insect branch of the arthropod race, perhaps the single most important adaptation that allowed them to eventually dominate every habitable ecosystem on Earth.

The small size of insects, certainly a problem for most other animals on land, gave them an advantage when trying to get into the air. They also got a boost from their highly adaptable appendages, which were able to create wings, probably from airfoil-like flaps that continued to evolve as the advantages of flight were passed from generation to generation.

Arthropods' tracheal breathing systems, evolved during the colonization of dry land, also proved very efficient at oxygenating the fast-moving muscles necessary for flight. Insects began to fly soon after their ancestors first came ashore, and when they did, their dominance of land began in earnest. Flight gave them a speed advantage over other ground-bound insects and animals of all kinds that was simply insurmountable. The ground speed of the fastest-running insect, for instance, is about 5.6 miles per hour; for a flying insect, about 35 miles per hour.

Once the air was alive with animals, it became a new food niche for others that were able to capitalize on meals hurtling through the sky. Spiders. These ground-bound arthropods figured out how to catch flies, grasshoppers, moths and their other airborne cousins in great nets spun across the sky. A spider web is one of the most marvelous structures on Earth, spun in patterns that have delighted humans forever and strong enough, if proportionate in size, to stop a flying 747 airliner. Flight also was the greatest step in the weaving of an ecological tapestry upon which all animal life on Earth now depends—the synergy between flowering plants and insects.

A vast majority of people consider it a high priority to minimize the extent of their interaction with the insect world. . . .But because insects are in many cases the chief architects of terrestrial ecosystems, they are also our principal partners in making a living on Earth. Without insects, there would be no oranges in Florida, no cotton in Mississippi, no cheese in Wisconsin, no peaches in Georgia and no potatoes in Idaho.

May R. Berenbaum, Entomologist,
Bugs in the System: Insects and Their Impact on Human Affairs

Arthropod statistics, past and present are staggering. Of the about one-and-a-half million or more described species of living animals, over one million of them are arthropods that together consume the greatest amounts and kinds of food on land and in the sea, and occupy the widest variety of habitats. Because of the entangled dependencies of all ecosystems, their success has also rendered most of the rest of the world's living things literally helpless without them. Many flowering plants, for instance, require insects for their survival, a biological collaboration of such importance that life as we know it would cease to exist without it. These two empires of plants and animals are intricately united, with insects consuming every part of the plant, and most flowering plants dependent on the insects for pollination and reproduction. Ultimately, plants owe bugs their lives because, like earthworms, insects turn the soil around their roots and decompose dead tissue into the nutrients for growth.

Insects and other land-dwelling modern arthropods are so important that some people speculate if all were to disappear, humanity could not last more than a few months and most of the other mammals, amphibians, and birds would crash to extinction at the same time. Without arthropods, dead vegetation would pile up and dry out, closing the nutrient cycles and forcing the end of living plants. As Earth adjusted itself to this new ecological formula without insects, fungi would explode into the unoccupied niches, but soon, they too would disappear as their own nutrient cycle ceased to function. The continents of Earth would return to about the same condition they were in before animals made landfall, covered by mats of bacteria and primitive plants.

Eventually, though, the arthropods might invade again to reclaim the land.

A SPIDER SPINS A WEB AND WAITS (OPPOSITE PAGE). □ FLYING INSECTS CARRY POLLEN FROM FLOWER TO FLOWER, AIDING POLLINATION AND ENSURING SEED PRODUCTION (ABOVE).

MOLLUSCS

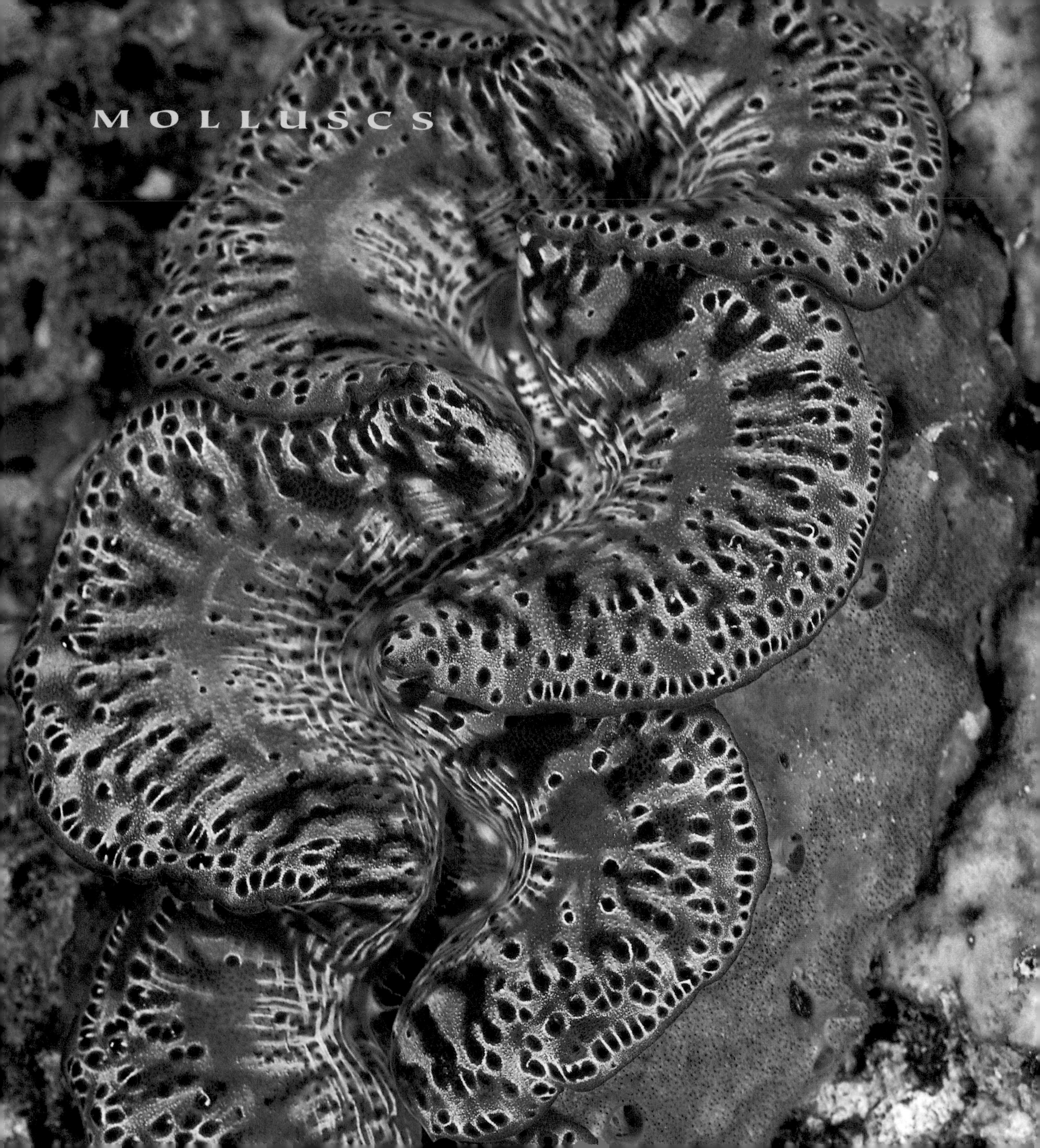

CHAPTER 6

ARMOR AND SPEED

The Survival Game

Alive and violated, they lay on their bed of ice.
Bivalves: the split bulb, and philandering sigh of ocean.

Seamus Heaney, Poet

Most of us encounter molluscs for the first time on our forks. With the exception of the occasional chocolate-covered grasshopper, a worm or two on a dare, and the always-delicious lobsters and crabs, molluscs are at the top of the invertebrate menu for humans. We eat tons of oysters, clams and mussels live, barbecued, and broiled with cheese and spinach. We devour queen conchs in chowder, scallops en brochette, escargot drenched in garlic butter, octopus on our sushi platters, abalone steaks, and deep-fried calamari to soak up the beer in sports bars. We hunt them, farm them and, for centuries, have treasured their delicate flavors. Special runners and pack trains were dispatched over the Alps from the Mediterranean to bring oysters to Roman legions in their inland garrisons of conquered Europe.

Our particularly intimate relationship with molluscs is due largely to their tasty, fleshy bodies, which have been an easy meal for millions of other animals in the food web since they first emerged during the Cambrian Explosion. With that kind of vulnerability, and because they couldn't outrun predators, early molluscs came up with armor to survive. The shells that make them easy to handle in a New England shore dinner gave molluscs an enormous advantage in playing the survival game over the millennia. Of the million or so named species of animals alive on Earth today, molluscs account for about 70,000. That's more than the 50,000 species in our tribe, the chordates, but far behind the 1.2 million species of arthropods. (The number of existing mollusc species is subject to rancorous debate, ranging from 50,000 to 110,000.) The adaptations that enabled molluscs to cope with day-to-day danger and threats of mass extinction were made possible by the most versatile body plan in the animal kingdom.

Death is the mother of beauty; hence from her, alone, shall come fulfillment of our dreams.

Wallace Stevens,
"Sunday Morning"

Earth has always been a dangerous place for animals. The climate swings between ice ages and searing heat, bringing cyclones, floods, and droughts. The oceans rise and fall, volcanoes and earthquakes fracture the land and seafloor, and even the continents themselves change shape and move around on the surface of the globe. Day by day, animals contend with the constant pressures of finding food and safe places to bear young. They must always be on guard against each other because most animals are part of the indifferent drama of prey and predator. These conditions drive the biological engine of evolution and force the adaptations upon which the survival of life depends.

And every so often, like loaded dice in a rigged craps game, destructive events of such enormous magnitude occur that all life is threatened. Over the 580-million-year history of multicellular animal life on Earth, these mass extinctions have been a skip in the heartbeat of the planet, and like the deadly outcomes of predator and prey, they are paradoxically essential to the existence of life. These sudden episodes of great mortal consequence come in varying shades of destruction, ranging from relatively minor die-offs that kill thousands of species to the five major events that have threatened the very existence of life on Earth.

The extinction at the end of the Cretaceous period claimed 50 percent of all species, including dinosaurs and the lesser known but equally dominant creatures called ammonites, the ancient molluscan relatives of clams, oysters, snails, mussels, squid, and octopus. The evidence of this stupendous catastrophe is indisputable. At many sites around Earth, there is a clear demarcation in the rocks below which lay the fossils of dinosaurs, ammonites and many of the other extinct species, and above which there are absolutely none. Zero. Those rocks can be precisely dated by measuring the amount of certain atoms that decay at a predictable rate so we know the exact time of their formation. There is no doubt that extinctions of massive proportions have occurred.

But why?

Peter Douglas Ward is a paleontologist known for his work on molluscs and, more recently, breakthroughs in our understanding of mass extinctions, the conditions that precede them, and the effects on the survivors. Ward and others cite a number of possible causes of these great die-offs,

GEOLOGICAL TIME LINE.

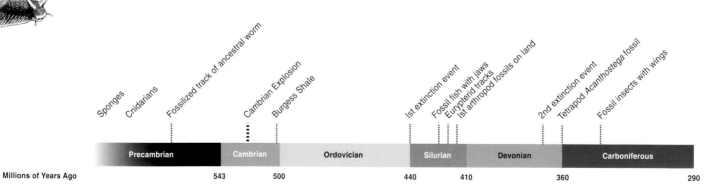

including a dramatic shift in climate; a sudden change in the composition of the atmosphere; volcanic global winter; changes in the fauna and flora that impacted major food webs; sudden drops in sea levels; a reversal in Earth's magnetic field that left portions of the globe exposed to solar wind and radiation from space; the explosion of a supernova near our solar system; and the impact of an asteroid or comet that plunged Earth into a period of fire, then darkness.

"Some theories have been fanciful, such as world-covering floods; others have been religious, citing God's will," Ward says of the explanations for extinction events. "Many of these have been crackpot ideas, while others emerge from great good humor. A tabloid headline blaming 'Big Game Hunters from Outer Space' and Gary Larson's view that cigarette smoking did in the dinosaurs are my two favorites." Very likely, Ward and other scientists believe, the catastrophic events that mark the great mass extinctions have had different causes at different times.

Because the number of species is currently in precipitous decline, Peter Ward and a growing number of paleontologists, biologists, and others tracing the course of life on Earth are becoming increasingly convinced that the planet is in the middle of the sixth great extinction event. The other five events: 1) 440 million years ago, 25 percent of the 450 families of animals living disappeared; 2) 370 million years ago a similar number of families went extinct; 3) 250 million years ago about 50 percent of the 400 families vanished, totaling up to 95 percent of all marine animals; 4) 200 million years ago 20 percent of the 300 families were wiped out; and 5) the celebrated extinction of the dinosaurs 65 million years ago, which also took 15 percent of the 650 marine families then alive, totaling about half of all animals.

It is difficult for us, as creatures capable of viewing life as precious, to look beyond the ultimate horrors of such declines in animal life and the violence of a culminating asteroid strike or other final catastrophic event that marks a mass extinction on the geologic calendar. The truth is, though, that such terrible times result in periods of equally dramatic bursts of creativity among the survivors. "Little evolution is likely to occur unless extinction has greatly shaken up—and diminished the numbers of—species already in existence," writes paleontologist Niles Eldridge. "Extinction is absolutely vital to the evolutionary process. When vast areas open, evolution is rampant. And it is

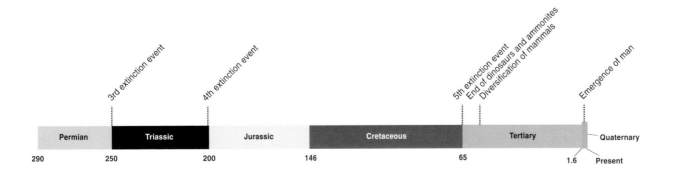

then that the large-scale transformations take place." When the dinosaurs were extinguished, for example, mammals found the way cleared for them to evolve into larger species in the land ecosystems and, eventually, lead to us.

The most remarkable part of the extinction story, though, is that of all the major body plans of animals that emerged during the Cambrian Explosion half-a-billion years ago, at least one representative of each has lived through all of the great and small mass extinctions. This stunning fact means that the genes that govern the creation of a particular body plan have also survived the long and dangerous trip into the present. When the last of the ammonites died with the dinosaurs, after a run of 300 hundred million years in the sea, for instance, enough other molluscs survived to continue that basic line of animal life, which began when that kind of animal diverged from its more primitive ancestors to begin its voyage through time.

MOLLUSCAN DIVERSITY: AN OCTOPUS, A CONCH SNAIL, NUDIBRANCHS (THIS PAGE, TOP TO BOTTOM), A THORNY OYSTER, KELLETT'S WHELK AND CHITONS (OPPOSITE, TOP TO BOTTOM).

All animals endure an arms race, all organisms endure an arms race. We have bacteria and anti-biotics, that's kind of an arms race. We have shells and shell-breaking predators, that's a classic arms race. Everything in biology really is an arms race.
Geerat Vermeij, Biologist

In so dangerous a place as Earth, however, the molluscs have always been good candidates for survival because they quickly diversify when the opportunity presents itself. The arms race of predator against prey has pushed the molluscs to grand new innovations, and extinctions have provided opportunities for rapid evolution within the group, resulting in an incredible range of variation on that single body plan.

The molluscs most familiar to us are the bivalves—the clams, oysters and mussels that show up so often on our plates. Each features a hinged pair of shells and a single, muscular foot with which it moves and burrows. Cephalopods (one of the great names in the animal kingdom, meaning "head-footed") are both the Einsteins and the Ferraris of molluscs—the squids, octopus, and cuttlefish. The hauntingly beautiful chambered nautilus is a cephalopod, too, an evocative living relative of the once-powerful ammonites. To those of us who prowl in tide pools, the members of the chiton branch of the family are also familiar as sluggish little creatures we call sea cradles that scrape algae from the rocks and are capable of clamping down like super-magnets. Chitons were most likely one of the earliest molluscs to emerge during the Cambrian and sport their armor in a simple arrangement of eight plates laid in an arch over their backs. About 650 species of chitons still exist.

And then there are the snails, known as gastropods ("stomach-footed," of course), those slow-moving creatures that have managed to colonize both land and sea, produce beautiful shells and torment gardeners who slaughter them by the zillions. Some gastropods, like slugs, have no shells and use stealth and prolific reproduction to survive even gardeners.

Shell-less slugs in the ocean, nudibranchs, are flamboyantly colored to advertise the fact that they taste bad to most other animals.

It takes quite a feat of the imagination to see how this diverse tribe sports the same body plan.

How can a clam, a snail, a slug, an octopus, and an extinct ammonite all be part of the same race? It's very easy to see if you know what to look for, because each has taken the same complement of body parts and crafted them for radically different tasks. The basic architecture of all the molluscs that drives their adaptability is a fleshy soft body. They get their name from a Latin word, *mollis*, which means "soft." Most molluscs have hard shells, but some have shed them in favor of speed.

All molluscs have a fold of tissue that covers the body, called a mantle, which is a spectacular piece of biological equipment that builds and shapes the incredible diversity of shells. Among bottom-dwelling species, the soft molluscan body has evolved a versatile foot, supported by internal fluids against which powerful muscles can work. In the abalone, the foot is a broad, flat, crawling platform also capable of clinging tenaciously to rocks, and in clams it becomes a tool for digging deep within the mud.

Also unique to molluscs is a special kind of tongue called a radula, which would be the envy of any woodworker or machinist. This raspy strap is backed by strong muscles and is coated with toothlike serrations used for scraping up food, lacerating flesh, or even for boring holes in other hard-shelled animals. Modern molluscs range in size from microscopic critters no bigger than a grain of sand to five-foot-wide clams to the biggest invertebrate of all, the giant squid that reaches lengths of 70 feet.

Pushed by constant threats from predators, the molluscan body plan showed its versatility early on in its history. The first molluscs in the early Cambrian were ground-bound like all the other animals on the seafloor, with shells to protect them against the dominant trilobites and other predators. The ability to manufacture that hard, versatile shell has been the key to the success of the molluscs, and we know them as aesthetic masters because of their magnificent shapes and colors, which turn beaches and knick-knack shell shops into art galleries.

Of course, molluscs do not create those beautiful shells for our adoring human eyes, but as part

SCANNING ELECTRON MICROSCOPIC VIEW OF RADULA WITH ROWS OF SHARP TEETH (THIS PAGE, TOP). UNDERSIDE OF AN ABALONE SHOWING THE RADULA IN ITS CIRCULAR MOUTH (BOTTOM). □ VENUS COMB SHELL (OPPOSITE PAGE, TOP LEFT). GEERAT VERMEIJ AS A CHILD (TOP RIGHT) AND AT WORK (BOTTOM).

of their strategy for survival. Geerat Vermeij, who probably knows more about molluscs and their shells than anyone alive, has been blind since he was a kid and has never seen one with his eyes. "I was first shown some shells from Florida by my fourth-grade teacher. I was prepared to like what I saw. Back in the Netherlands, I had already grown fond of shells. A successful day at the beach meant a good haul of cockles, wedge shells, and razor clams. I was overwhelmed with their beauty," Vermeij remembers of those Florida shells. "You know, they had such smooth interiors and such rather nicely sculptured and shaped exteriors. The contrasts were beautiful, the ribbing was even and the whole thing just struck me as a real work of art. And I never turned back. That was for me the beginning of a career, really. My teacher had not only given my hands an unforgettable aesthetic treat, but she aroused in me a lasting curiosity about things unknown."

Now, in his lab at the University of California at Davis and at field sites all over the world, Geerat Vermeij sees molluscs very clearly, interpreting the ancient history of these remarkable survivors and their elegant armor with his fingers and the power of his intellect. He marvels at molluscs and their ability to survive in nature's arms race. "I think that a body plan can be pushed incredibly far," Vermeij says, turning a spectacular, whorled shell in his hands. "When you think about molluscs, they're everything from slow, tanklike snails, to very, very fast squid that have large brains and very, very good eyes. There is virtually nothing that you can't do with a molluscan body plan. It may take evolutionary time to get there, but you can get there. And, well, for me, the most amazing things about molluscs is their sheer beauty. The diversity on a simple theme of spiral growth, how much you can do with a simple theme like spiral growth. They show all these wonderful variations, it's kind of like listening to Bach all your life."

More than a hundred years ago, another mollusc interpreter named Canon Moseley gave a mathematical account of the spiral forms of snail shells in one of the classics of natural history. To a mathematician, the shape of a shell can be read in Moseley's series of numbers and symbols, but in words . . . the surface of any shell, whether shaped like a disc or a turban, revolves around a fixed axis as a closed curve, which continually increases in dimensions as the animal grows larger. In molluscs shaped like disks, such as clams, oysters and scallops, to name a few, the shell grows in a very open spiral. In turban-shaped shells, like those of garden snails, conchs, or so many of the other collectable beach treasures, it grows in a helical spiral along the axis of the shell.

Molluscs can't do math, or express themselves in complex relational terms, so how do they make shells? Their secret lies in that unique mollusc body part called the mantle, which is a filmy sheath covering its soft inner flesh like a cape. The mantle secretes the layers of shell from its outside edge, building with ingredients it takes into its body as it feeds and breathes. The mantle lays down a matrix of protein that determines the shape of the shell as it grows, and calcium carbonate from the

sea water hardens within it. The spectacular colors unique to each species of mollusc have evolved as a result of the combination of each particular animal's diet, habitat, and internal chemistry. Shells grow in stages from the leading edge outward, and each spurt of construction is usually marked by a clear line in the finished shell, giving them their distinctive and graceful patterns. Many mollusc shells sport spikes and protrusions, which have evolved to make them more difficult to break or eat.

Most molluscs are either male or female, shedding their sperm and eggs into the sea, or mating and laying eggs. Some, including land snails, nudibranchs and some clams, are hermaphroditic, although they usually need another mollusc to get the job done, since self-fertilization is rare. The fertilized eggs of many marine snails and bivalves hatch to become drifting larvae, called veligers. Squid, octopus and the other cephalopods lay eggs, which, like the eggs of terrestrial molluscs, develop directly into juveniles without passing through a larval phase.

Gastropods—the marine snails that make all those beautiful shells—perform a unique bit of molluscan gymnastics as they develop from veligers to adults. Nobody is quite sure why this peculiar process evolved, but in an architectural modification called torsion, a snail flips its shell 180 degrees, twisting its body in relation to its head and foot. Molluscs, and all animals except sponges, cnidarians and flatworms, form two openings as they grow from fertilized egg into either larvae or adults, one

THE INTERNAL STRUCTURE OF A SHELL OVERLAPS TO PREVENT BREAKAGE (ABOVE). ☐ MOLLUSCAN SHELL DIVERSITY (RIGHT).

opening becoming the mouth, the other the anus. As they grow, these two openings usually remain as widely separated as possible on the adult body. With torsion, however, a mollusc throws a figure-eight twist into this arrangement, after which both the mouth and anus are pretty much situated at its head. This doesn't sound like much of a good deal, unless you happen to live in a shell with only one opening and an elastic body plan that can do just about anything to survive, including defecating on your own head. Despite such a compromise, torsion gives snails more room in their shells into which they can withdraw. It also brings their gills up front where they can more effectively extract oxygen. And they've devised ways of creating currents within their shells that sweep their waste away from their heads.

Life in a hard shell has enormous advantages if you are soft and slow. To us, molluscan armor has become a source of *objets d'art*, but in the vicious reality of the Cambrian ocean where its inventors were scrambling to stay alive, it was pure animal genius. The creation of shells was a matter of following the essential law of natural selection, which allowed those members of a particular mollusc

generation whose mantles secreted thicker coverings to survive and reproduce. Those that were eaten because they were the softest among them did not pass on the grow-a-thicker-shell trait to their offspring. Molluscs, and most of the survivors that emerged in the early going of animal life on Earth, were quick studies. Given millions and millions of years, they refined their original ability to produce shells because they were driven to do so by the pressures of snapping claws and predators' jaws.

In nature, though, every dog has its day, as we say, and eventually molluscs refined their adaptive powers to find a way to grow jaws of their own, start moving around in the water, and become fierce predators. It was yet another demonstration of the grand versatility of their body plan. They evolved a way to chamber their shells and control buoyancy, and a new branch of their race—the cephalopods—rose off the bottom. This was an enormous step forward in the arms race for survival, and very bad news for trilobites and other animals that could only get off the bottom in brief spurts, but were not truly mobile in the water column. Those first swimming molluscs were the ancestors of the modern nautiloids, squid, cuttlefish and octopus, and they revolutionized the uneasy relationships between the hunter and the hunted. The swimming molluscs quickly dominated the marine food web, and their descendants have spread to every ocean on the globe, terrifying lesser creatures, including ourselves.

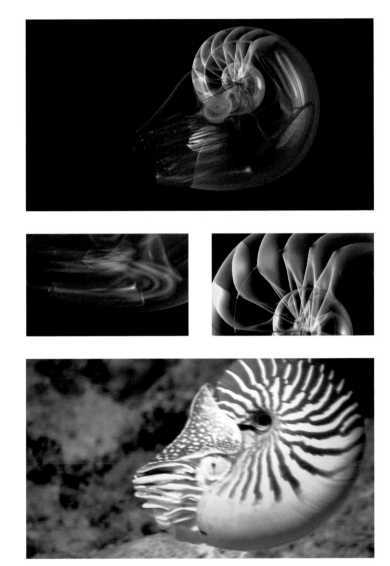

> *The history of life can be great theater. . . . The development of the gas- and liquid-filled chamber in the shell liberated the nautiloids from the sea bottom and set in motion an evolutionary history that is still unfolding today.*
> Peter Ward, Paleontologist,
> *On Methuselah's Trail: Living Fossils and the Great Extinctions*

Getting off the bottom of the ocean while wearing a hard shell was no simple piece of biological business. Nature selected this path to survival and dominance, of course again demonstrating the ability of the molluscan body plan to evolve a winning solution to a problem. This happened about 500 million years ago, and the shell of the earliest swimming mollusc was an elongated cone. As the animal grew, it sealed earlier shell growth into chambers with a tube connecting each compartment from the front, where the animal lived, to the smallest chamber in the rear, which it had created first. This connecting tube, called a siphuncle (from the word siphon) contained tissues that emptied the water from its chambers, leaving a gas-filled shell that was light in the water and could even counter the added weight of a growing animal.

You can see the blueprint for molluscan buoyancy the next time you're in a trinket shop at the beach where the split shell of that living fossil, the chambered nautilus, is always a star attraction. This stupendous development meant that the animal had the architecture for controlling its buoyancy. The rest of its anatomy evolved with wonderful possibilities.

GASTROPOD SNAILS (OPPOSITE PAGE, TOP). *ORTHOCERUS* WAS AN EARLY CHAMBERED-SHELLED CEPHALOPOD (MIDDLE, BOTTOM). □ AN ANIMATED STILL OF A NAUTILUS SHOWING ITS INTERNAL STRUCTURE (THIS PAGE, TOP). WATER IS PUMPED FOR JET PROPULSION (MIDDLE LEFT). THE SIPHUNCLE RUNS THROUGH ALL THE CHAMBERS (MIDDLE RIGHT). A LIVING NAUTILUS (BOTTOM).

By modifying their muscles to eject water at velocities great enough to push them through the water, the swimming molluscs became masters of jet propulsion. Add to that the evolution of big, complex eyes, a hard beak capable of crushing just about anything else that crawls or swims, and a forest of suckered tentacles on its head, and you have one of the most powerful killing machines of the ancient ocean.

Even when fishes and giant reptiles showed up in the sea, nautiloids and their cousins the ammonites held their own. They even survived the greatest extinction event of all time at the end of the Permian, 250 million years ago. From their beginnings as cone-shaped monsters that could be six feet long, the nautiloids morphed into hundreds of species, most of which curled their shells into discs. These relatives, the ammonites, grew to gigantic size. The largest ever found measured over nine feet in diameter, and by the end of their run as a branch of the cephalopod class of molluscs, ammonites had bent their shells into every imaginable shape, including very weird saxophonelike creatures that lived while the last of the dinosaurs were running things on land. When the asteroid ended the Cretaceous 65 million years ago, the ammonites went with them, so nobody has ever seen one alive. One ancient, patient nautiloid lives on, though, celebrated on the cover of this book as an example of durability and wonder contained in a single animal body plan.

. . . Starbuck still gazing at the agitated waters where it had sunk, with a wild voice exclaimed—
'Almost rather had I seen Moby Dick and fought him, than to have seen thee, thou white ghost!'
'What was it, Sir?' said Flask.
'The great live squid, which they say, few whale ships ever beheld, and returned to their
ports to tell of it.'

Herman Melville,
Moby Dick

And the story of the wildly adaptable molluscan body just gets better. After the cephalopods lifted themselves and their shells off the bottom of the oceans, speed became a key to survival for some of them, and so nature selected for that trait when fishes became predators. Like modern airplanes, power systems improved, bodies grew more streamlined, and in some species, the increased speed meant a heavy defensive shell became less and less important. Eventually, the exterior shell became smaller, moved inside the body and all but vanished. That left only a vestigial strip of flexible tissue

VARIOUS FOSSIL AMMONITES (THIS PAGE, TOP LEFT). RED OCTOPUS (BOTTOM). □ SCHOOL OF SQUID (OPPOSITE PAGE).

running lengthways in their otherwise soft bodies. This feather-shaped remnant of a hard exterior shell, called a pen, has a plasticlike texture and lies beneath the enveloping mantle, now morphed into a thick, tubular muscle.

Enter the squids, torpedo-shaped, jet-propelled predators that range in size from a few inches to sixty-foot giants that can move at speeds up to 20 miles per hour and fear no creature on the planet. Squid, and their cousins the octopus, are the most highly organized invertebrates that ever lived. A squid has three hearts to pump oxygenated blood through a complex closed-capillary system not unlike our own. Its

nervous system is centralized and also highly complex, with part of it fused into a brain center between its eyes, which are very large in proportion to the rest of its body. These are real eyes, too, capable of forming images. Remarkably like our own, they're built on the same principle as a camera, which consists of a dark chamber into which light is admitted only through a lens. (The eyes of humans and squid are examples of convergent evolution, evidence that strikingly similar anatomical structures have evolved more than once in different animals.) Some squid add the ability to produce disorienting clouds of ink or bioluminescence to their defensive and offensive arsenals, and all have hard beaks, some of which are capable of doing immense damage to anything else that swims. Their tentacles and arms are adorned with powerful suckers that have appeared in the nightmares of many a blue-water sailor. And as if mobile armor and speed are not enough of a triumph of adaptation, molluscs also came up with intelligence. Octopuses and cuttlefish think, learn and react to their environments in ways surpassed only by vertebrates. But they are still molluscs, using the same body plan as clams, snails, limpets, and the rest of their relatives, running the evolutionary race against destruction and transforming themselves to survive in a very dangerous place.

Think about all that the next time you sit down to a plate of linguini con calamari.

One wonders why some animals last and others go extinct. In a humble clam, a beautifully evolved squid, or a nautilus, we can tell that certainly they all evolved from a single, common ancestor. A mollusc.

Peter Ward, Paleontologist

ECHINODERMS

CHAPTER 7

GIVE ME FIVE

An Ultimate Animal?

An insightful seventeenth-century Frenchman could have been talking about echinoderms when he said, "It is well to comprehend clearly that there are some things that are absolutely incomprehensible." So might a twentieth-century marine biologist when he said, "If there are animals from another planet already here, they're probably starfish."

Ralph and Mildred Buchsbaum and John and Vicki Pearse,
Living Invertebrates

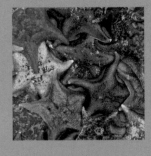

The evolution of animal life has not followed the rules of good theater by revealing its themes, characters and conflicts in clear stages, leading to an inevitable climax. If it had, the story of body plans would have proceeded directly from sponges through cnidarians and flatworms to the explosion of annelids, arthropods, molluscs, and chordates, stopping there with a fanfare of trumpets, drums and applause. After all, that story line produced heads, brains, eyes, intricate systems for active living, and a variety of bilateral shapes that met the heroic task of surviving for hundreds of millions of years. But at the end of what should have been the final act in the drama of animal complexity, the echinoderms waltzed in from the wings, took center stage and said, "No. Hold on. That's not the only way to evolve. Let's try it again."

Even if you accept the notion that nature has no obligation to make sense, echinoderms are enigmas that evolved like no other animals on Earth. Their curtain went up over a half-billion years ago, along with those of all the other basic blueprints for building an animal, but they tell an entirely different evolutionary story. Most animal life set off in the direction of bilateral bodies, heads, central nervous systems and brains, adapting complex tissues, organs, and sensory systems to live in the sea, on land, and in the air. This powerful assembly of body components made great sense for hunting, surviving the attacks of other animals, and continuing to adapt to constantly changing environmental conditions. Those lines of animals evolved into the billion or so species that have called Earth home ever since. At some point, though, one group diverged so radically it hardly seems to belong to the same world of the others. These are the echinoderms. Their name means "spiny-skinned," but their unique portrayal of life is about much more than that.

In a fast-paced world, the echinoderms chose to live in the slow lane. Their bodies seem to be little more than skeletons and water. They don't use large muscles working on large body parts like other animals. Instead they move on hundreds of tiny, water-filled tube feet operated by a hydraulic system that can't produce high-speed movement. Their skeleton is made of tiny calcareous plates, and their five-sided bodies interact with the world equally from all directions, without a head to lead

the way. This low-energy body doesn't require lots of fuel. How did such an alternative lifestyle get created and sustained by the evolutionary forces that drove most of the animals of the Cambrian towards an active bilateral approach to life?

The first sure evidence we have of the presence of these non-conforming interlopers comes from fossils that formed in the early Cambrian Explosion, about 525 million years ago, and they don't look much like their modern descendents. The primitive echinoderms lived on the seafloor and moved slowly, if at all. Some of the traits of their race are present in these early fossils, including the calcite skeleton and the hydraulic water-vascular system powering their tube feet. Yet it appears that evolutionary experimentation was taking place with body symmetry. Here are bilateral forms as well as three-sided and four-sided bodies, and even globular shapes without symmetry. It could be that the earliest stages of the echinoderm way of life were bilateral, but for some reason that approach didn't work well within the constraints of their other, already established body parts.

The five-sided, or radially pentamerous, body now sported by most modern echinoderms doesn't show up until a few million years later in a group of extremely weird creatures called crinoids. These characters, easily mistaken for plants, were clearly five-sided forms splaying open five arms to catch tiny food drifting by in the water. Five-sided symmetry was a good strategy for living a filter-feeding life attached to the bottom. If you're stuck on the seafloor in one place, why not meet the environment equally from all directions? This way of life was extraordinarily successful. Some forms even got themselves off the bottom and into better position for filter-feeding by growing stalks. Many of them became giants, rising as high as 70 feet off the seafloor, not by swimming but like the first human aviators who rode as lookouts in tethered balloons above battlefields. With bilateral symmetry catching on all around them, with fronts and backs and heads and tails proliferating, with brains and eyes beginning to evolve, the early bilateral echinoderms just might have been outdone by other bilateral animals. So they began doing things differently.

Echinoderms have such an unusual body plan, you might think that it was just an experiment that didn't go anywhere, but that's quite wrong.

Andrew Smith, Paleontologist

Since they abandoned their early bilateral ancestry, echinoderms have been on a kind of evolutionary side trip. Imagine coming up with a new way to build a land vehicle that is elementally different than the cars, trucks, and sports utility vehicles clogging today's highways. To achieve the same magnitude of divergence that echinoderms managed from the basic architecture of all other animals, you'd have to dispense with wheels, hoods, trunks, and most of the components of engines. You'd have to abandon the concepts that a vehicle moves in forward or reverse gear, that it is propelled by a single power source, and even that a driver is required to drive it. You would have to abandon all of the highways, bridges, tunnels, and speed limits that co-evolved with the design of cars, trucks and sports utility vehicles. Whatever you come up with in a radically new vehicle will require an entirely different environment for moving around. The evolution of vehicles for common use on highways has already produced a workable result, so yours would be very, very different and not at all intuitive or logical. And perhaps not even possible.

It is worth noting that the spiny-skinned echinoderm revolutionaries *do* possess the basic traits of animalness. They are made of many cells doing specialized work but interacting to produce a functioning whole; they are the product of the fertilization of a large egg by a smaller sperm; and from this single cell they transform themselves in a highly organized way into an adult body. The regulatory Hox genes governing that transformation are amazingly similar in all bilateral animals.

But what about the echinoderms? As it turns out, they also have some of those same Hox genes, but in their bodies they direct the formation of an entirely different shape. Modern echinoderms still produce larvae that are bilaterally symmetrical, more evidence that leads some investigators to conclude that they must have risen from an early bilateral ancestor.

Being different, however, seems to have worked out just fine for the echinoderms. About 6,500 named species of their race still inhabit Planet Earth, not a giant tribe compared to the 1.2 million kinds of arthropods or 70,000 molluscs, but one spread throughout all the world's oceans. Echinoderms are almost exclusively bottom-dwellers, although many have larvae that spend some time drifting in the water column before abandoning that lifestyle for the security of adulthood on the seafloor.

Five major adaptations of the echinoderm body plan exist today, commonly known as sea stars, sea urchins (and their close cousins, sand dollars), brittle stars and basket stars, sea cucumbers, which constitute the largest biomass on the deep sea bottom, and crinoids, which include feather stars and sea lillies. You can group animals any way you want to— eyes, or no eyes; legs or no legs; or by color, or aroma or whatever. But if you are interested in defining the relationships of animals within groups through time, a system based on anatomical characteristics works best,

BAT STAR, PURPLE URCHIN, CROWN OF THORNS STAR, CRINOID, TROPICAL URCHIN (OPPOSITE PAGE, TOP TO BOTTOM).
☐ FOSSIL CRINOIDS (THIS PAGE, TOP), BILATERAL ECHINODERM LARVA (MIDDLE), CRINOID (BOTTOM).

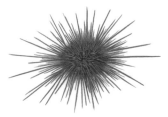

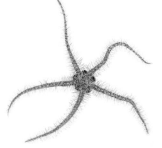

and we've been using it since the seventeenth century when the Swedish scientist Linnaeus laid it out.

Since echinoderms emerged over 500 million years ago, at least 14 major adaptations, or classes, of that body plan have risen and died off, in addition to the five classes currently alive. These were huge groups of related animals within which diverse families, genera, and species evolved, and then vanished. Many other classes of animals in other phyla have also gone extinct, but never so many in proportion to the number of echinoderms that survived into the present. The expression of so many extinct classes within one phylum appearing, thriving, and becoming extinct is a microcosmic representation of the evolutionary vigor that went into creating the separate phyla of animals.

They are exciting. They are wonderful. It takes some time and some patience to understand them, but the more you look, the neater they get. You really have to watch, they really don't give away their secrets easily.

Gail Kaaialii, Biologist

Just about everyone has seen an echinoderm, probably one of those exotic, colorful creatures in a tide pool, on a dock piling, or in an aquarium. Some are so obviously evocative of celestial objects that we have named them sea stars (class Asteroidea). (Really, a star in the night sky is a ball of fire like our sun, but when we look at it from afar it seems to radiate arms of light. Somewhere along the line, perhaps because it's easy to draw, we came up with the five-pointed star as an icon.)

Imagine picking up one of those sea stars, or recall your first real and daring encounter with so strange a creature in an aquarium touch tank. When you pick up the sea star, you notice that it quickly becomes rigid in your hands, a hardened star. Just moments ago, it was flowing like rubber over the rocks in the tank. When threatened, many echinoderms have the ability to freeze themselves and remain so for a long time by changing the physical state of the collagen connecting the tiny plates of their skeletons. In this way, they can hold a position without expending much energy.

Look at the animal in your hand. Like most adult sea stars, its arms radiate in multiples of five from a central hub. This rule of five can be broken, though, so you might see sea stars with anywhere from four to fifty arms. One giant, *Pycnopodia*, may have more than 20 arms and can grow as large as a manhole cover. Other echinoderms also follow the rule of five. The body of a sea urchin is like an origami sea star, with its arms folded toward the tips of each other to form a ball, but you can still see the five distinct radials, just as you can in its close relative, the sand dollar. A sea cucumber looks very little like a star and more like a

THE FIVE ECHINODERM CLASSES: SEA CUCUMBER, BRITTLE STAR, CRINOID, SEA URCHIN AND SEA STAR.

cucumber, but it is really made up of five distinct sections organized into an elongated tube, an adaptation to a bilateral way of life without losing the rule of five. And if you look closely at crinoids and brittle stars, you'll see their obedience to pentamerous radial symmetry as well.

A sea star has no head or tail, front or rear, but most do have a top and bottom. Feel the sea star with your fingers, and sense its textures and contours. It's kind of rough and pebbly, right? Many sea stars, and especially sea urchins, use spines as part of their defense armament, which can nick you, sometimes very painfully. But your sea star, let's say a common tidepool character known as *Pisaster*, is harmless to a big, intelligent, bipedal carnivore like yourself. The top surface of its body is covered by a thin skin, which is a living epidermis and can extract oxygen and dissolved nutrients from the water. This skin covers an internal skeleton of calcareous plates arranged in a specific pattern. This unique internal skeleton keeps the animal together and gives it shape. The plates are held together by tough connective tissue and small muscles that allow it to flex around rocks, pilings and crannies like an alien ballerina when it isn't frozen in your hand.

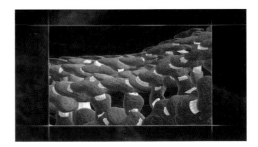

Now turn your sea star over. Like all animals, an echinoderm has to eat to live. See that opening in the middle? That's its mouth. An echinoderm's digestive system is quite simple. Beginning with that mouth on the bottom side

Sea star, urchin, and sea cucumber share the same body plan (left). A sea star's skeleton (top right) is held together by connective tissue and muscles (bottom right).

and ending with an anus on the top, it is basically a stomach with branches extending down the arms for digestion.

To eat, echinoderms hunt prey, graze on plants and organic debris, scavenge on the bodies of dead animals, catch plankton, or simply absorb nutrients through their skin from the sea. That sea star from the touch tank is a ferocious, slow-motion predator when it comes to hunting down and eating mussels that can't move at all and are at the top of the star's menu. The sea star senses the mussel with receptors at the tips of its arms, moves to cover it, and clamps down with some of its tube feet. Then it slips part of its stomach through a crack between the shells of the doomed mollusc, and settles down to digest its meal by liquefying the meat inside.

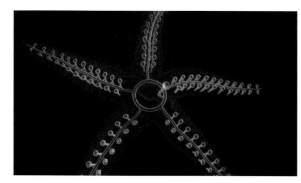

Are you still looking at the underside of the sea star? That's the really creepy part, figuratively and literally. You are looking at the marvelous, totally unique echinoderm hydraulic system for getting around. Tube feet. Once you get over the heebie-jeebies from seeing what appears to be a wriggling infestation of worms, look more closely and you can see an array of tubes that together are highly coordinated, very efficient, water-driven feet. They work like this: Sea stars, and most echinoderms, are plumbed with a water vascular system that radiates as five canals down each arm from a central ring canal encircling the mouth. Rows of tube feet extend out from each canal, projecting through the body wall. Sea water for the system is pumped in by cilia through a central sieve plate that looks a little bit like the drain in a hot tub, and distributed to the canals and tube feet. When you look at those wriggling tube feet, you're looking at the extremities of the water vascular system, and one of the marvels of the echinoderm way of life.

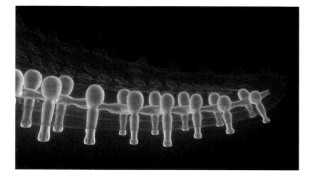

The sea star in your hand has thousands of tube feet, each of which it can extend by increasing the pressure inside a little muscular bulb filled with water. This forces more water into the tube foot to extend it. Muscles in the tiny tube foot itself move it from side to side or retract it back up toward the arm. Tube feet are used for locomotion, adhering to

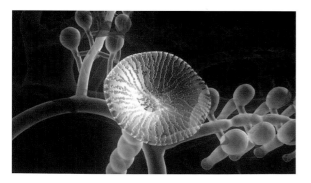

A SEA STAR'S WATER VASCULAR SYSTEM WITH TUBE FEET, BULBS AND THE SIEVE PLATE (LEFT, FROM TOP). A SEA STAR'S STOMACH EXTENDING FROM MOUTH (ABOVE). □ NERVOUS SYSTEM WITH NERVES RUNNING DOWN ARMS AND A CENTRAL RING.

rocks, pilings and other objects, breathing, burrowing, moving food around, sensory perception, or performing a combination of tasks. Some tube feet at the ends of a sea star's arms have evolved tiny primitive eyespots that can sense light. By coordinating the pressure in their vascular system and the muscular movement of all their tube feet, sea stars can move equally well in any direction, climb things, and clamp themselves tightly to stay put.

And they do it all without a brain. Using your X-ray vision (remember, you're imagining this) look inside the sea star. Instead of a brain, an echinoderm has a set of nerves that radiate in five directions from a central nerve ring encircling its mouth. This nervous system relays impulses from light, tactile, and chemical sensors to the nerve ring, which acts as a kind of relay station rather than a central processing organ to coordinate the movement of its arms, feeding tools, and tube feet. If the radial nerve in one of a sea star's arms is cut, that arm will continue to move but in a different direction than that of the other arms still connected to the central nerve ring. If the central nerve ring itself is cut in half, the sea star may literally pull itself apart in two directions with the power of its tube feet that are no longer working together. However, this presents no real problem for an animal that can regenerate lost parts to make itself whole again.

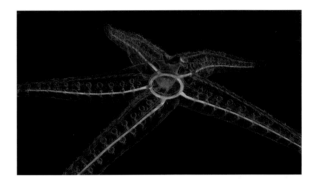

One of the really fascinating things about echinoderms is that they don't seem to grow old. They can live forever. The only thing that kills a sea star is physical harm or disease. Some sea stars can regenerate their whole body from one part of the ray. Just take a little ray, and cut it off, and the whole thing will grow out. How do they do that?

John Pearse, Biologist

Most echinoderms can regenerate lost parts easily, and even create two or more animals from a single one. In a classic story of invertebrate revenge, fishermen pestered by infestations of sea stars routinely chopped them up and threw the parts back into the sea, wondering all the while why their problem kept getting worse and worse. Like all animal traits, the ability to regenerate evolved according to the rules of natural selection. Since most sea stars live in rocky, shallow water subject to the violent combination of surf, surge, and boulders, those that could perform this remarkable act of self-preservation prevailed to become ancestors of future generations that would carry the ability to regenerate. And when a predator has a grip on one of your arms, what better way to escape than by simply leaving it behind?

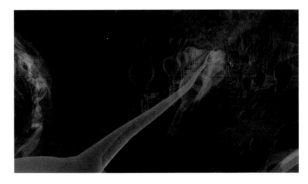

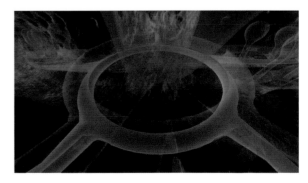

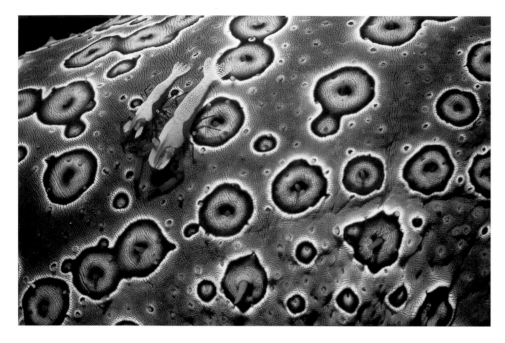

When an arm breaks off, the wound seals itself and a new arm begins to grow. Sometimes, the severed part can grow an entire new animal if it has a piece of the central disc. One sea star actually uses this regenerative ability as its main method of reproducing and creating successive generations—deliberately casting off an arm that will grow into a new animal. Echinoderms also reproduce sexually, although males and females look very much alike. They release eggs and sperm into the water and fertilization occurs externally, producing bilaterally symmetrical larvae that actually swim, propelled by minute cilia. These larvae metamorphose from their bilateral beginnings into the typical, radially symmetrical adults, possibly in a replay of their ancient origins.

When you investigate the lives of other organisms and you see that each is interesting and wonderful, you start to realize that the question of whether anybody is better becomes moot, irrelevant. It's not something you'd want to ask. You start to ask how do the others work—they're so neat.
Gail Kaaialii, Biologist

Through most of our history of looking at echinoderms, it's been easy to think of these denizens of rocks, pilings, kelp forests and seafloors as brightly colored clumps that somehow make a living but don't do much in the way of interacting with each other or the rest of the ocean's flora and fauna. Only in the past half-century, really, have we been admitted into the world of five-sided living with scuba tanks and underwater movie cameras to witness firsthand the remarkable world of these silent, spiny-skinned creatures that took off in their own direction on the evolutionary trail. We now know that echinoderms invented as successful a set of solutions for meeting the challenges of life as any other race of animals. We have not only picked up sea stars in aquarium touch tanks to marvel at their wildly different body plan, but with the help of time-lapse photography, watched them digest a mussel, or witnessed voracious, grazing sea urchins laying waste to entire kelp forests. Urchins, like all echinoderms, live on a time scale that is very much slower than our own, one much more suited to life as a five-sided radial animal than one with a head and someplace to go in a hurry. Sea stars and urchins look like odd, colorful but inert objects sitting on the bottom, but if their action is sped up, they clearly demonstrate hunting, defensive, and social behavior as vigorous as that of any other animal.

The smallest echinoderms are brittle stars, graceful ballerinas of the sea that defend themselves by shedding their arms. They make up for their size and fragility by covering vast expanses of the seafloor like a furry, waving carpet, delighting many a diver. With deep-ocean submersibles, we have also spied on the secret life of sea cucumbers and found that they absolutely rule the bottom of the abyss. Without brains, eyes, or speed,

these vacuum cleaners of the sea make up 90 percent of the living biomass on the deep ocean floor, which means they have essentially conquered two-thirds of the surface area of the planet. Sea cucumbers have taken an elongated version of the pentamerous, radially symmetrical echinoderm body plan, and adapted the tube feet at one end to shovel nutrient-rich sand into their mouths. No head? No brain? No tail? No problem.

Echinoderms are the bohemians of the animal kingdom, but there is nothing about their history to suggest that they made a mistake in taking the road less traveled.

It's simply wrong to think that evolution favors one path over another. . . . It's much better to think of the tree of life as a bush with lots of branches coming out in all directions.
Andrew Smith, Paleontologist

SHRIMP LIVING ON SEA CUCUMBER (OPPOSITE PAGE, TOP). SIEVE PLATE ON SEA STAR'S CENTRAL DISC (BOTTOM). □
BRITTLE STARS IN KELP HOLDFAST (THIS PAGE, TOP). SPINY SURFACE OF SEA STAR (BOTTOM).

CHORDATES

CHAPTER 8

THE VOYAGE TO US

Bones, Brawn and Brains

Humanity is exalted, not because we are so far above every other living creature, but because knowing them well elevates the very concept of life.

E. O. Wilson,
Biophilia

The main entrance to the Smithsonian Institution's National Museum of Natural History is a grand rotunda with colonnades rising to a domed ceiling that magnifies the constant din of voices and the footsteps clattering on the marble floors below. It takes a few minutes to get used to the scale of the enormous hall and echoing chatter from rivers of meandering children tended by parents and teachers, and the rest of humanity that gathers there every day to marvel at the story of life. Once you settle into the sound and the crowd, though, the sight of a giant elephant at the center of the rotunda banishes the sensory overload. The animal, long dead but perfectly lifelike in a natural setting, is so much bigger than you could have imagined that it hardly seems of this world. This is a beast, a great beast, not some gray cartoon creature or decorated Pasha's mount or laborer on a circus lot. This is a rough, fierce, darkish brute that looks dangerous and quick, its bright white tusks hinting of the unthinkable damage they could inflict, its huge head easily 12 feet from the floor. It is an African bush elephant, you learn from reading the provenance panel mounted on the rail around the display, the largest modern living land animal. It weighs 15,000 pounds and exhibits intelligence, nurturing instincts, and complex social behavior in the wild. As do we.

Radiating from the rotunda, the halls of the Smithsonian contain mounts or models of thousands of the creatures with whom we share Planet Earth. A great blue whale, the largest animal that ever lived, hangs overhead in one of the galleries. In another, a warren of display cases holds a collection of birds so vast you could spend most of a morning just there. In other halls of similar magnitude you'll find fishes, reptiles, amphibians, and more mammals. Because all of the animals alive today comprise less than 10 percent of all the animals that ever lived, the ancient fossilized remains of many members of these families of extinct creatures also tell their stories, including the legendary monsters in the Hall of Dinosaurs, the hit attraction of the Smithsonian for over a century. Since the rush to excavate their fossils began in the middle of the nineteenth century, so many people have seen dinosaurs in museums, movies, and books that they are as familiar to us as our pets. Many children can say *Tyrannosaurus rex* soon after they learn to talk.

Fewer than five percent of all the animals that ever lived on Earth have backbones, but we humans remain quite chauvinistic about our own kind. Natural history museum collections tend to

be heavy on chordates, particularly the branch to which humans and most familiar animals belong, the vertebrates. If you ask most people to name five animals, you'll almost certainly get a list of creatures with backbones and nothing else. Try it. Cat. Dog. Snake. Monkey. Horse. Some invertebrates, though, are represented in the exhibit cases and dioramas, including insects, spiders, and mollusc shells. And one entire wing of the Smithsonian is devoted to life in the ancient oceans, which were home to swarms of the spineless creatures that inhabit the first seven chapters of this book.

Those ancient seas were also home to some of our earliest chordate relatives. They didn't look much like us, or the giant beast in the rotunda, and in fact didn't even have real backbones. To find one, you take a right from the elephant and pass under an archway into a dimly lit gallery where you can hear the sloshing sounds of the sea piped through unseen speakers. You walk past great slabs of crinoids, cases of trilobites, a collection of Walcott's beautiful Burgess Shale fossils, a stunning diorama of a Permian reef, and a wall of swimming ammonites and marine reptiles. And there, tucked into a corner to the right of the reef, in a shrinelike case that looks a little bit like a glass-covered draftsman's table, is a black slab of rock in which you can see the outlines of a wormlike creature named *Pikaia gracilens*.

Charles Walcott himself brought the first specimen of *Pikaia* to the Smithsonian from his legendary quarry. It was one of thousands of messages from the Cambrian Explosion embedded in the dark shale of the Canadian Rockies that would eventually change the way we know ourselves, and all other animals on Earth. Walcott thought *Pikaia* was a worm, one of many he found there on the high flank of Mt. Burgess, in British Columbia, Canada, along with the arthropods which seemed to dominate that epoch in the sea. His mistaken identification of this tiny creature in the shale is quite understandable. It clearly has a head with a pair of hornlike appendages, a tail, and a flattened, elongated body in between. Even today, almost anybody would look at *Pikaia* and think 'worm.' And so it remained for 70

VERTEBRATES ARE THE MOST FAMILIAR CHORDATE GROUP. SOME EXAMPLES: KING COBRA, LEAF MONKEYS, GALAPAGOS TORTOISES, TIGERS AND GIRAFFES (THIS PAGE, TOP TO BOTTOM). □ ANIMATION STILL OF *PIKAIA* SWIMMING IN THE CAMBRIAN SEAS (OPPOSITE PAGE, TOP). FOSSIL *PIKAIA* WITH NOTOCHORD BARELY VISIBLE ABOVE THE SEGMENTED MUSCLES (BOTTOM).

years until modern paleontological tools and new understandings of the relationships among the animals combined to tell us its real story. And our own story. As so often happens in the world of paleontology, new fossils have since been found that are even older than our friend *Pikaia*. We now know that in these first chordate fossils lie the beginnings of the chordate body plan, expressed as a rod, or notochord, running the length of the backside of the body. When you look closely at *Pikaia* through the glass of the display case in the Smithsonian, you can actually see it, a wispy filament etched in the rock, not much thicker than one of the hairs on your head. You can also see a pattern left by the segmented muscles that is the harbinger of the creatures who would not only dominate Earth but some day leave it to explore space.

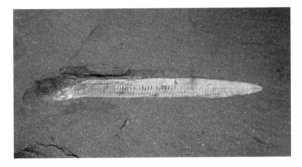

And so the chordates arrived, soon to be followed by their powerful and enormously successful descendants, vertebrates with real backbones. Their animalness—multicellularity, nerves, a head, brain, and bilateral symmetry—is the legacy of the sponges, cnidarians, and ancient flatworm-like animals, each of which pioneered pieces of the architecture that led to complex animals like *Pikaia*, and us. From the Cambrian, by which time those basic traits had been established, all of the other body plans emerged that animals still use today. So in addition to the three pioneer groups continuing to evolve, all of the other groups of animals, including the annelids, arthropods, molluscs, and chordates, began their bilateral journeys through time into the present. (Echinoderms, of course, began at the same time but navigated into the future in their own way with their radially pentamerous interpretation of animalness.)

And though *Pikaia* could not walk, talk or write this book, its body plan contained the traits shared by every chordate and vertebrate that ever lived. Our chordate body plan is easily defined by three common features. Each of us has a single hollow nerve bundle running along our backs, and that stiff rod, called a notochord, containing fluid-filled cells sheathed in fibrous tissue that we saw etched in the fossil, *Pikaia*. In some living chordates, the notochord remains flexible, but in vertebrates, it is incorporated into the structure of the spine as discs between the vertebrae. The third trait shared by all chordates is the presence, at some stage of life, of gill slits in the throat. Our own gill slits close up and disappear after we leave our watery beginnings as embryos. At some point in our lives, all of us chordates have bilaterally symmetrical bodies and have one-way guts that run from a mouth to an anus, and a well-developed central body cavity in which our complex internal organs are suspended. We reproduce sexually from the union of a large egg and a much smaller sperm, and in most of us, the sexes are separate. We can see hints of ourselves in the fossil of the extinct *Pikaia gracilen*s if we know what to look for, but how could evolution possibly have crafted our big complex bodies from those humble beginnings? Surprisingly, there is an animal alive today in which we also find clues to our ancestry.

It's a long way from amphioxus, it's a long way to us,
It's a long way from amphioxus, to the meanest human cuss,
Goodbye fins and gill slits, hello skin and hair,
It's a long, long way from amphioxus, but we get here from there.

An Anonymous Paleontologist
(To the tune of "It's a Long Way to Tipperary")

Amphioxus, which looks a lot like *Pikaia* and is still around, gives us a wonderful window through which we can observe the behavior and genetic makeup of early chordates. They thrive in the sand of tidal flats with essentially the same anatomical equipment as *Pikaia*. The shallows of Tampa Bay, for instance, are alive with these tiny, translucent, fishlike little creatures that have no eyes, ears, or jaws but which carry one of the earliest chapters in the story of the evolution of animals like us.

By studying how an amphioxus develops from an embryo to an adult, investigators like biologist Linda Holland trace the path taken through the millennia by our chordate ancestors. She has spent countless hours dredging amphioxus embryos from the shallows and trying to piece together their story back in her lab at Scripps Institution of Oceanography. "Amphioxus is giving us a lot of insight into how we evolved," Holland says. "We're interested in it because back in the fossil record there

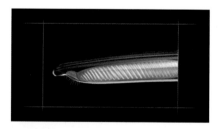

were little organisms that looked very much like this and we can begin to reconstruct a scenario for exactly how the vertebrates evolved from little invertebrate creatures like these." Like *Pikaia*, amphioxus has a nerve cord, and underneath that a notochord that acts as a stiffening rod against which its segmented muscles work as it wiggles in the sand. It has gill slits for feeding on tiny particles of food from the surrounding water. This animal is definitely a chordate.

The big clue to the origins and relationships of these primitive chordates and

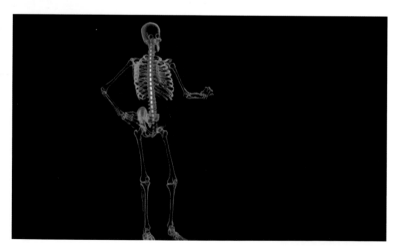

ourselves, though, is not in the gross anatomy but in the way its regulatory genes direct the formation of the adult body. During their first 48 hours of existence, amphioxus embryos grow from pinpoint-sized eggs to actual swimming larvae. Holland isolated and studied the genes of amphioxus at this stage of development, compared them with those of vertebrates, and found remarkable similarities. "One of the most exciting moments was when I got the very first pattern of where a gene was turned on in the nerve cord of an amphioxus," Holland says.

"And it was the same as in the mouse nerve cord or in the human or the chick nerve cord." In a simple living

descendant, we see how the genes that made the first chordate could have also created our bodies. This means that an animal very much like amphioxus should be pasted on the first page of the vertebrate family album, because the regulatory genes that direct our own development have a similar function in that little creature.

> *Tracing the roots and branches of one's family tree can be embarrassing enough, but anyone sufficiently curious to extend the search phylum-wide should be prepared to acknowledge some even more unlikely relatives.*
>
> Ralph and Mildred Buchsbaum and John and Vicki Pearse,
> *Living Invertebrates*

The phylum Chordata includes about 50,000 species, most of them vertebrates like us—fishes, amphibians, reptiles, birds, mammals. About three percent of chordates are invertebrates, though, including amphioxus and an extremely unlikely branch called tunicates. Some tunicates look for all the world like red, yellow, purple, orange, and green jelly smeared on rocks on the seafloor or in tide pools. Their name means "enclosed in a cape." We also call them sea squirts, but unless you know better, you would look at a tunicate and call it a sponge. Nobody could blame you. But when you take a look at a tunicate larva through a microscope, its startling connection to us and the rest of the chordate clan is obvious. The first thing likely to pop into your mind when you look at a tunicate larva is 'tadpole'. That miniscule creature has the same features—a nerve cord, notochord, and gill slits—as the rest of us chordates until it attaches itself to the seafloor or some other object and metamorphoses into a complex adult body.

Other invertebrate chordates, commonly called salps, skip the larval stage and instead

ANIMATION STILL OF AMPHIOXUS SHOWING ORANGE GILL SLITS, AND THE NOTOCHORD A FAINT YELLOW JUST BELOW THE RED NERVE CORD (OPPOSITE PAGE, TOP). THE DISCS BETWEEN THE VERTEBRAE IN THE HUMAN SKELETON ARE ALL THAT REMAIN OF THE NOTOCHORD (BOTTOM). □ COLORFUL SESSILE TROPICAL TUNICATES (THIS PAGE, TOP). AN INDIVIDUAL SALP (MIDDLE RIGHT) AND A SALP CHAIN (BOTTOM).

of attaching themselves forever to one place, slowly glide through the ocean filtering their meals. Some drift through the ocean like giant bubble chains. Swarms of individual salps can be gigantic, reaching densities of up to 25,000 individual animals per cubic yard and covering vast expanses of the ocean. Fishermen who blunder into a large infestation are tormented by the chains and clumps of salps coming up in their nets and on their hooks that are almost impossible to remove. Not many of those fishermen, of course, suspect that they are closely related to the goo that is driving them crazy.

> *Einstein was a fish. So were Mozart, Virginia Woolf, and Thomas Jefferson. Bruce Springsteen is a fish, and so are Ray Troll, Bill Clinton, and Camille Paglia.*
>
> Brad Matsen and Ray Troll,
> *Planet Ocean: A Story of Life, the Sea*
> *and Dancing to the Fossil Record*

Nor do many trawlers and trollers tortured by the salps suspect how closely they are related to the fish they are trying to catch. The truth is, those fishermen themselves are more fish than not, because some ancient fish, about 520 million years ago, established the vertebrate line. The very first vertebrates crossed the bridge from the spineless world by adapting the basic chordate body plan to become the most dominant predators in the history of the animal kingdom. *Pikaia* and amphioxus are chordates, but they and other animals like them do not have backbones, skulls, or complex sense organs and nervous systems. So what had to happen to the body of an animal like amphioxus to transform it into a fish?

"Vertebrates really became big, dominant animals by getting extra genes," says Linda Holland,

THE DEVELOPMENT OF JAWS WAS IMPORTANT IN THE EVOLUTION OF VERTEBRATES. MANY LIVING FISHES HAVE LARGE JAWS.

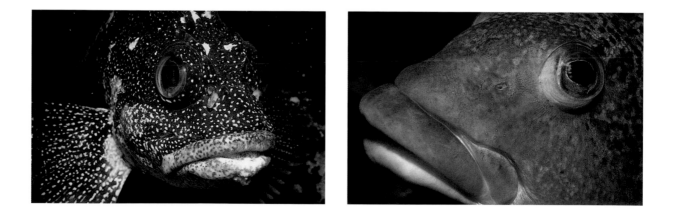

whose amphioxus larvae are the 'before' in a before-and-after picture. "An amphioxus-like organism has relatively few genes, about the same number as your typical worm, your typical ant, your typical fly. But vertebrates have done something rather special. They've taken this basic number of genes, and instead of inventing a lot of extremely new genes, they've taken them and simply duplicated them." Flies have about 10,000 distinct genes; annelid worms about 13,000. Mice and humans, though, have about 40,000. "Vertebrates duplicated the genes of their simpler ancestors not once, but twice. Then they changed them just a little bit, and suddenly, you have four times as many genes to link together to make brand-new structures."

By quadrupling the amount of genetic information available for choreographing their development, fish could evolve big, complex bodies and forever change the world. Some of those duplicated genes directed the evolution of a true backbone, the key trait of the vertebrate group whose members would eventually dominate land, sea and air. Others were responsible for an innovation that was the evolutionary equal of the backbone, a group of specialized cells that began building body parts, like skulls and jaws. These fantastic cells develop near that nerve cord to form a very powerful component of vertebrates called the neural crest. In animals like amphioxus, this neural crest hardly forms at all, and neither does a backbone. In true vertebrates, the neural crest has enough genetic horsepower from all those duplicated genes to send cells off on journeys through time and space to create such complex structures as skulls, and, eventually, jaws, teeth and parts of the nervous system.

Working with this powerful genetic tool kit, fish became the first true vertebrates. The first fish were jawless vacuum cleaners, probably looking very much like enlarged, muscular amphioxus. They had inherited the basic architecture of all chordates, and then, about 480 million years ago, real bones arrived and started fishes and vertebrates on the road to global dominance and space travel. The first bonelike structures were external plates that covered the fish like armor, but over time internal bones evolved. Bone is an extraordinary building material, composed of living tissue that can grow and respond to increased weight and stress.

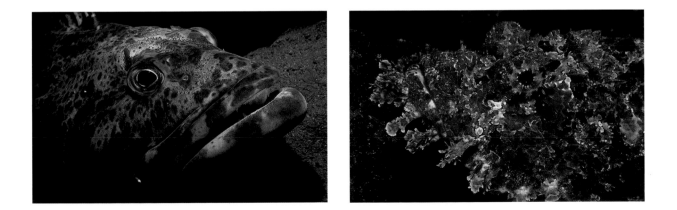

"Once jaws evolved," Linda Holland says, "chordates could switch . . . to becoming predators, eating other animals and increasing their body size, and so we went from something as small as an amphioxus to something as big as an elephant, a cow, a horse and a lion." Fishes were also the smartest animals around in the Paleozoic seas, with big brains that could process enormous amounts of sensory information from eyes, sense organs and lateral lines that detect motion, and translate all that information into complex behavior. They quickly became the top predators in the oceans. More than half of all vertebrates that ever lived have been fishes, including members of two extinct groups: the armored fishes and the spiny-ray sharks; and four living groups: the jawless fishes; the sharks, rays and chimeras; the lobefin fishes; and the ray-fin fishes, which are spectacularly diverse in today's oceans, lakes and rivers.

The greatest step in vertebrate evolution is undoubtedly the transition from aqueous gill-respiring fishes to air-breathing, walking, land animals.

John A. Long, Vertebrate Paleontologist,
The Rise of Fishes

Life 400 million years ago in the Devonian Period, also known as the Age of Fishes, was driven by the enormous evolutionary pressures of jaws, teeth, slashing tails and fins, and speed that produced an explosion of diversity. Among those ancient fishes, a group called lobefins emerged. They had paired fins with the pattern of bones that laid the foundation for the arms and legs of the chordates to follow. They had hip and shoulder muscles to move those fins. And they had primitive lungs that could supplement their gills during times of stress. While they still lived in the water, such animals were well suited to eventually evolve into creatures that could make the leap to life on land and become four-legged terrestrial animals.

Powerful evidence that limbs evolved while tetrapods were still water-bound comes from a fossil discovered in the mountains of Greenland, in 1987, by Jenny Clack, a paleontologist at Cambridge University's Museum of Zoology. That evidence, a fossil dating back 360 million years ago, emerged from tons of rocks that Clack brought back to her lab and painstakingly investigated

RAY CRUISING WITH ITS WINGLIKE PECTORAL FINS (OPPOSITE PAGE). □ MODERN FISH DIVERSITY (THIS PAGE, CLOCKWISE FROM TOP LEFT): LIONFISH, JUVENILE COWFISH, HORN SHARK AND QUEEN ANGELFISH.

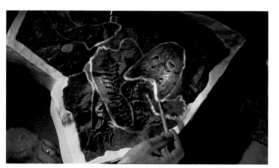

for clues to the origins of land vertebrates. Among the hundreds of specimens she uncovered was the most complete Devonian tetrapod ever found, a creature named *Acanthostega* that she affectionately calls 'Boris'.

"It's every paleontologist's dream to find a transitional form," says Jenny Clack, "something that falls between two groups that we are familiar with, that sort of links them both in terms of anatomy and lifestyle. When we fetched Boris from the field, most of the surface of the fossil was covered by rock that had to be dug out bit-by-bit. We suspected that we had gotten something exciting, because we could see lumps in the rock suggesting there was more to the specimen than met the eye. And we could see the cross cracks suggesting that there were things inside waiting to come out."

Sealed in its tomb of ancient stone, Boris is a fishlike animal with limbs. "This is a specimen of an animal that could be described as a missing link, except that we have one so it's not missing," Clack says, holding her fossil. "It's a transitional form between animals with fins that we'd call fish, and animals with legs, with fingers and toes on the end that we call tetrapods. We are tetrapods. Like us, Boris has a skull attached to a vertebral column, which goes through this S-bend, like this, and goes off into a tail. And then there are forelimbs and some digits here. If you look at the back of the skull here, there's a rodlike structure. And when we turn the skull over you can see some more of those. These are gill bars, and the significant point about these is that they are grooved here, and that means that there was an artery running up that groove, feeding the gills blood, so that the blood could get oxygen. And that suggests that the animal was still using gills to breathe. It would have had lungs, anyway, because most of these early fish did. But this animal was using both gills and the lungs, unlike later tetrapods, where the gills were lost." Clack has also found the tracks of an early tetrapod left in the fossilized mud of the shallows of an ancient shore, which further completes the puzzle of vertebrate landfall. These footprints are evidence of the process of evolution, and though the species disappeared about 360 million years ago, its descendants have walked the Earth since those first tentative footsteps left their marks on the surface of our planet.

And so the genetic fire that began with those earliest chordates like *Pikaia* burned through fishes

Jenny Clack (top). The tetrapod fossil, Boris, with the backbone twisted to the left of the head and a leg with eight digits outlined (bottom).

and amphibians and up onto the land where reptiles became the first fully terrestrial keepers of the chordate flame. Their limbs proved to be perfect for getting around on land, and though snakes abandoned them to better exploit their particular niche, forelimbs and hindlimbs became the rule for survival ashore. The early reptiles also protected their young in the absence of the watery womb of the sea by evolving liquid-filled eggs with tough shells, which meant they didn't have to return to the ocean to reproduce.

For over 50 million years, amphibians and reptiles were the sole vertebrate terranauts, sharing islands and continents with insects, worms, snails, and the other animals that had preceded them onto land. Then, at some point before the mass extinction that ended the Permian Period 250 million years ago, the reptiles were driven by evolutionary pressure to send off a branch that would become the ancestor of mammals. After the extinction event, the radiation of reptiles on the land took off in several directions at once, with dinosaurs appearing and other reptiles returning to the sea where they evolved into the giant mosasaurs, plesiosaurs, and ichthyosaurs that eventually became, for a time, apex predators of the world's oceans. On land, the dinosaurs retained much of their reptilian ancestry, continuing to lay eggs, but the mammals evolved further terrestrial adaptations, grew hair, and began to bear live young. Mammals survived millions of years of ecological change in part by remaining small and resource-efficient, much like modern rodents. Somehow they managed to survive the asteroid strike and mass extinction that killed off the dinosaurs and marine reptiles. When the dust settled, mammals and birds exploded into the vacated environmental niches on land, in the sea, and in the air, and the chordate story continued, now with the familiar characters that to most people are *real* animals.

There are about 4,500 species of mammals alive today. We are homeotherms. Warm-blooded creatures. We have four-chambered hearts, with complete double circulation that keeps oxygen-rich blood separate from oxygen-depleted blood. We have hair on our skin at some stage of our lives, and we nourish our young with milk secreted by the mammary glands of females. We reproduce sexually, and the fertilized egg develops inside the female, nourished in most mammals by a special organ, the placenta. Many of us have complex and differentiated teeth. We eat plants and other animals, including other mammals.

JENNY CLACK AND SARAH FINNEY EXAMINE FOSSILIZED TETRAPOD TRACKS (TOP). ANIMATION STILL OF THE TETRAPOD WALKING IN SHALLOW WATER (BOTTOM).

Some of the members of our class include rodents (squirrels, mice, and porcupines); insectivores (hedgehogs, shrews, and moles); chiropterans (bats); carnivores (dogs, cats, and bears); ungulates (deer, horses, cows, sheep, and goats); pinnipeds (seals and sea lions); cetaceans (whales and dolphins); and primates (lemurs, monkeys, apes, and human beings).

As primates, we share a common ancestry that we can clearly see in the great apes and bonobos whose features and behavior are indisputable evidence of the anatomy we share. Our connection to all other animals is less obvious, but we can see it if we fearlessly transport ourselves upstream in the genetic river through our vertebrate, chordate, and animal ancestors who don't look much like us at all. We *are* there in that sponge in a tide pool, in the anemone stuck to a dock piling, in the earthworm on Charles Darwin's lawn, and they are in us. And although we write books, leave Earth and land on the Moon, peer through telescopes looking for the origins of the universe, sing the great duet in the first act of *La Boheme*, and weigh the abstractions of good and evil in a world once governed only by the impulses of predator and prey, it is clear that we are neither more, nor less, than any other animal that ever lived.

And a little later on, your friend goes out to the Moon. And now he looks back and he sees Earth, not as something big, where he can see the beautiful details, but now he sees Earth as a small thing out there. It is so small and so fragile and such a precious little spot that you can block it out with your thumb, and you realize that on that small spot, that little blue and white thing, is everything that means anything to you—all of history and music and poetry and art and death and birth and love, tears, joy, games, all of it on that little spot out there that you can cover with your thumb. And you realize that there's something new there, that the relationship is no longer what it was.

Russell Schweikart, Astronaut,
Earth's Answer: An Exploration of Planetary Culture

A FEW OF THE MAMMALIAN ORDERS (OPPOSITE PAGE, CLOCKWISE FROM TOP LEFT): SEAL (PINNIPED), ZEBRA (UNGULATE), FRUIT BAT (CHIROPTERAN), BONOBO (PRIMATE), TREESHREW (INSECTIVORE) AND LION (CARNIVORE), AND THIS PAGE, BOTTLENOSE DOLPHIN (CETACEAN) AND MOUSE (RODENT).

BIBLIOGRAPHY

BIBLIOGRAPHY AND FURTHER READING

Barrington, E. J. W. *Invertebrate Structure and Function*, Houghton Mifflin, 1967.

Berenbaum, May R. *Bugs in the System: Insects and Their Impact on Human Affairs*, Addison-Wesley Publishing, 1995.

Briggs, Derek E. G., D. H. Erwin, F. J. Collier. *The Fossils of the Burgess Shale*, Smithsonian Institution Press, 1994.

Brusca, Richard C. and Gary J. *Invertebrates*, Sinauer Associates, Inc., 1990.

Buchsbaum, Ralph and Mildred, John and Vicki Pearse. *Animals Without Backbones* (Third Edition), University of Chicago Press, 1987.

_____. *Living Invertebrates*, Boxwood Press, 1987.

Crick, Francis. *Life Itself, Its Origin and Nature*, Simon and Schuster, 1981.

Darwin, Charles. *The Origin of Species by Means of Natural Selection or the Preservation of Favoured Races in the Struggle for Life*, (Reprint), New American Library, 1958.

Dawkins, Richard. *River Out of Eden, A Darwinian View of Life*, HarperCollins, 1995.

Eldridge, Niles. *Life Pulse: Episodes from the Story of the Fossil Record*, Facts on File, 1987.

Fortey, R. *Life: A Natural History of the First Four Billion Years of Life on Earth*, Random House, 1998.

Gerhart, John and Marc Kirschner. *Cells, Embryos, and Evolution: Toward a Cellular and Developmental Understanding of Phenotypic Variation and Evolutionary Adaptability*, Blackwell Science, 1997.

Gould, Stephen J. *Wonderful Life: The Burgess Shale and the Nature of History*, W. W. Norton & Company, 1989.

Haeckel, Ernst. *Art Forms in Nature*, Dover Publications, Inc, 1974.

Langstroth, L. and Libby Langstroth. *A Living Bay: The Underwater World of Monterey Bay*, The University of California Press/Monterey Bay Aquarium, 2000.

Lenhoff, Sylvia G. and Howard M. *Hydra and the Birth of Experimental Biology, 1744: Abraham Trembley's Mémoires Concerning the Natural History of a Type of Freshwater Polyp with Arms Shaped Like Horns*, Boxwood Press, 1986.

Margulis, Lynn and Karlene V. Schwartz. *Five Kingdoms: An Illustrated Guide to the Phyla of Life on Earth* (Second Edition), W.H. Freeman and Company, 1988.

Margulis, Lynn and Dorion Sagan. *What is Life?* Simon and Schuster, 1995.

Matsen, Brad and Ray Troll. *Planet Ocean: A Story of Life, the Sea, and Dancing to the Fossil Record*, Ten Speed Press, 1994.

Morris, Simon C. *The Crucible of Creation: The Burgess Shale and the Rise of Animals*, Oxford University Press, 1998.

Palmer, Douglas. *Atlas of the Prehistoric World*, Discovery Books, 1999.

Raff, Rudolf A. *The Shape of Life: Genes, Development, and the Evolution of Animal Form*, University of Chicago Press, 1996.

Ricketts, E. F., Jack Calvin with Joel W. Hedgpeth, revised by David W. Phillips. *Between Pacific Tides*, Fifth Edition, Stanford University Press, 1985.

Robison, B. and J. Connor. *The Deep Sea*, Monterey Bay Aquarium Foundation, 1999.

Schopf, J. William. *Major Events in the History of Life*, Jones and Bartlett, 1992.

Stanley, Steven M. *The New Evolutionary Timetable: Fossils, Genes and the Origin of Species*, Basic Books, 1981.

Vermeij, Geerat J. *Evolution and Escalation: an Ecological History of Life*, Princeton University Press, 1987.

_____. *Privileged Hands: A Scientific Life*, W. H. Freeman and Company, 1997.

Vogel, Steven. *Life's Devices: The Physical World of Animals and Plants*, Princeton University Press, 1988.

Ward, Peter Douglas. *In Search of Nautilus: Three Centuries of Scientific Adventures in the Deep Pacific to Capture a Prehistoric-Living-Fossil*, Simon & Schuster Inc., 1988.

_____. *The End of Evolution*, Bantam Books, 1994.

_____. *On Methuselah's Trail: Living Fossils and the Great Extinctions*, W.H. Freeman, 1992.

Watson, James D. *The Double Helix*, Atheneum, 1968.

Wells, Martin. *Lower Animals*, World University Library/McGraw Hill, 1968.

Zimmer, Carl. *At the Water's Edge: Macroevolution and the Transformation of Life*, The Free Press, 1998.

PHYLOGENETIC TREE OF LIFE

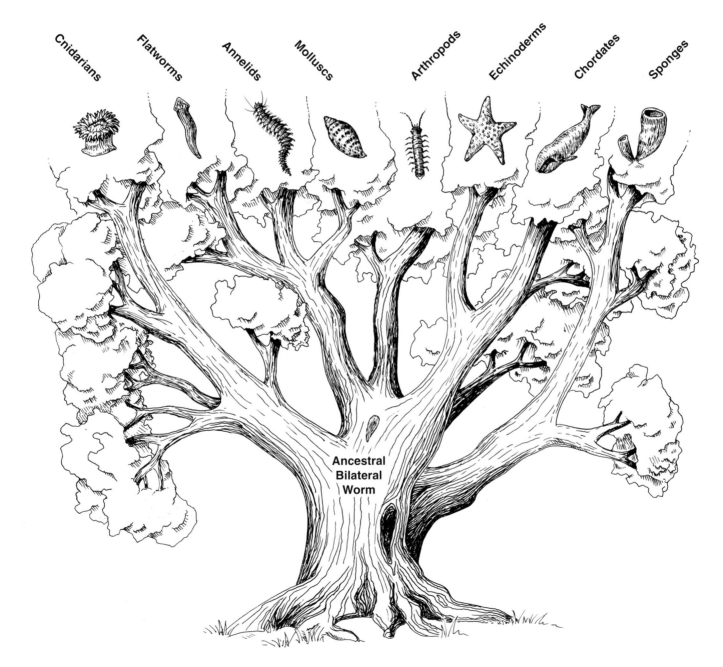

Cnidarians
Flatworms
Annelids
Molluscs
Arthropods
Echinoderms
Chordates
Sponges

Ancestral Bilateral Worm

INDEX

INDEX

abalone, *82*

ammonites, 78, 80, *88*

amphibians, 111

amphioxus, 104–105, *104*, 106–107

anemones. *See* sea anemones

angelfish, queen, *109*

animals
    bilateral bodies, *38*, 39, *44*, 45
    common characteristics, 3, 19, 93
    emergence onto land, 64, 71–73
    environmental adaptation, 5–6
    genetic relationships, 16–17,
        45, 50
    heads, 38–39, *38*
    humans and, 3, 113
    mass extinctions, 78, 79, 88, 111
    origin of, 3–4
    Praya as longest animal, 32
    self differentiation by, 15
    single-celled. *See* protists
    species' diversity by natural
        selection, 5–6
    sponge as ancestor to, 7–9, 15–
        16, 17
    See also classification

annelids, *6*, *48*, 55
    body characteristics, 55, *57*, 58
    burrowing, 61
    Canadia spinosa, 55
    carbon recycling role, 61
    heads, *56*
    hydrostatic skeleton, 59
    polychaetes, 59, *59*
    segmentation, *57*, 59
    tube worm, *59*, *60*
    tube-dwelling, tentacles and
        fans, *60*, 61

Anomalocaris, 53–54, 58

Aristotle, 4, 9

arthropods, 7, *64*
    centipedes, *64*, 71
    exoskeletons, 68–69
    eyes, *68*
    fossils, 51, *51*, 53, 71, 72
    jointed appendages, 67–68, *67*, 69

marine, *65*

millipedes, 71

molting, 68–69, *68*

name origin, 67

number of species, 75, 77

plants as food for, 73–74

respiratory systems, 67, 69–71,
    73, 74

size limits, 68–69

trilobites, 51, 64, *64*, 86

See also centipedes; crabs;
    insects; lobsters; scorpions;
    spiders

Aysheaia (arthropod fossil), *53*

Australia, Great Barrier Reef, 13, 27

bacteria, in currents of sponges, *14*

bat, fruit, *112*

behavior, 15

Berenbaum, May R., 68, 74

Bilharz, Theodor, 35

bilharziasis, 35, 43

blood flukes, 43

body plans, 7, 56, 103
    bilateral symmetry, *38*, 39, *44*,
        45, 92–93, 103
    five-sided symmetry, 92, 94–95
    mass extinction survivals, 80
    See also classification; *under*
        names of specific animals

bonobo, *112*

"Boris" (tetrapod fossil), 110, *110*

Braddy, Simon, 71

British Columbia, Mount Burgess,
    51, *52*, 102

brittle stars, 99, *99*
    See also echinoderms

Buchsbaum, Mildred, 39, 43, 64, 91,
    105

Buchsbaum, Ralph, 39, 43, 64, 91, 105

Burgess Shale, 51, 102
    See also Walcott's Quarry

California, Inyo Mountains, 37

Cambrian Explosion, 51, 52, 54, 102

echinoderm emergence during,
    92

explanations for, 57–58, 61

mollusc emergence during, 77

See also geological eras

Canadia spinosa (annelid), 55

cells
    amebocytes, 11
    choanocytes, *10*, 11
    choanoflagellates, 10, *10*
    cnidoblasts, 26
    epithelial, 11
    neoblasts, 42
    porocytes, 11
    specialization, 10, 11

centipedes, *64*, 71

cephalopods, 81, *86*, 88–89
    cuttlefish, 89
    eyes, 89
    intelligence, 89
    nautilus, 87, *87*
    nautilus' internal structure, *87*
    octopus, *80*, *88*, 89
    Orthocerus, *87*
    See also molluscs; squid

chitons, *80*, 81
    number of species, 81

choanoflagellates, 10, *10*

chordates, 7, *7*, 102
    amphioxus, 104–105, *104*, 107
    body plan, 103
    gill slits, 103, *104*
    limbs, 109–110
    nerve cord, 104, *104*, 105, 107
    neural crest, 107
    notochord, 103, *104*
    number of species, 77, 105
    regulatory genes in, 104–105
    reproduction, 103
    salps, 105–106, *105*
    tunicates, 105, *105*
    See also fish; vertebrates

Clack, Jenny, 109–110, *110*, *111*

classification, 3–4, 5
    of humans, 4

flatworm, *46*
squid nervous system, 89
New Zealand flatworms, earthworm
killing by, 35–36, *36*
nudibranchs, *80*

oceans, underwater exploration,
63–64
octopus, *80*, 89
intelligence, 89
red, *88*
Orthocerus (cephalopod), *86*
oyster, *80*

parasites, 35, 43–44
Pearse, John, 39, 43, 64, 91, 97, 105
Pearse, Vicki, 39, 43, 64, 91, 105
phyla *See* classification
Pikaia gracilens, 7, 102–104, *103*, 106
planarian (flatworm), 41, 42
plants
as food for first land animals,
73–74
synergy with insects, 74–75
Platyhelminthes. *See* flatworms
polyps, 20, 22, 23
medusae evolution from, 27–29
reproduction, 27
strobilation, 29
See also cnidarians
Praya (cnidarian), 32
Proterospongia, 10, *10*
protists, 10, *10*
in currents of sponges, *14*

Raff, Rudolf A., 55, 56, 57
ray, *108*
reefs, 12–13, 27
regeneration of body parts, 20, 41–42,
97–98
reptiles, 111

salps, 105–106, *105*
Schiller, Joseph, 21
Schistosoma spp. (flatworm), 35, 43
schistosomiasis, 35
Schweikart, Russell, 113
scientific method, 20, 30, 49–50
experimental biology, 21

scorpions, *64*, 71
See also arthropods
Scotland, New Zealand flatworms in,
35–36
Scott, Matt, 44, *44*, 45
sea anemones, *18*, 21, 23
body parts, *22*
clone wars, 27
Metridium extending upward, *24*
muscles, 24–25
nematocysts in, 26, 27, *27*
reproduction, 27
skeletons, 24–25
See also cnidarians
sea cucumbers, *94*, 98, 99
body plan, 94–95, *95*
See also echinoderms
sea squirts. *See* tunicates
sea stars, *90*, *91*, *92*, 94, *94*, 98, 99
body plan, 94, 95, *95*
central disk, *98*
defense behavior, 94
digestive system, 95–96, *96*
muscles, 95, *95*
mussels as prey for, 96
nervous system, *96*, 97
Pisaster, 95
Pycnopodia, 94
skeletons, 95, *95*
skin, 95
spines, 95, *99*
tube feet, 96–97, *96*
vascular system, 96, *96*
sea urchins, *92*, *94*, 98–99
body plan, 94, 95, *95*
See also echinoderms
seal, *112*
segmentation
of annelids, *57*, 59
of arthropods, 67
shark, horn, *109*
Shear, William, 63, 72–73, *72*
shells
chambered, 87, *87*
crustacean, 68, 69
mathematical descriptions, 84
mollusc, 82–84, *83*, *84*, *85*, 87
natural selection and, 86
See also molluscs

shrimps, *64*, 98
banded coral shrimp, *70*
in sponge, 12
See also arthropods
skeletons
exoskeletons, 68, 69
human, *104*
hydrostatic, 24–25, 59
sea star, *93*
Smith, Andrew, 92, 99
Smithsonian Institution, 7, 52, 101,
102
snails, 86
snake, king cobra, *102*
Sogin, Mitch, 1, 15, 16–17, *16*
spiders, *64*, 71, 74, *74*
webs, 74, *74*
See also arthropods
sponge spicules, 12, 15
sponges, 6, *8–9*, 19
as ancestral animal, 7–9, 15–16,
17
behavior, 15
body plan, *9*, 10, 11
cell functions in, 10, 11
cells in the currents of, *14*
choanocyte cells, 10, *14*
classification, 4, 9–10
feeding, 14
glass, skeleton of, *12*
growing in coral, *8*
mobility, 15, 19
number of species, 9
porosity, *2*, *3*
Proterospongia as ancestor, 10, *10*
as reef builders, 12, 13
regeneration of body parts, 41
reproduction, 14–15
spicules, 12
spicules as Japanese wedding
gifts, 12
water pumping by, 10, 13–14, *13*
squid, 89, *89*
mobility, 89
nervous system, 89
three hearts, 89
See also molluscs
starfish. *See* sea stars
Steinbeck, John, 35

Stevens, Wallace, 77
Stomphia (sea anemone), 32

tapeworms
    fish tapeworm, parasitic cycle, 44
    heads, *43*
    record length, 35
    reproductive segments, *43*
    See also flatworms
Texas, Guadalupe Mountains, 13
tiger, *102*
tortoise, Galapagos, *102*
toxins, cnidarian, 25, 26
treeshrew, *112*
trematodes, 43
Trembley, Abraham, 19, 20–21, *20*, 30
trilobites, 51, 64, *64*, 67, 86

Troll, Ray, 106
tunicates, 105, *105*

urchins. *See* sea urchins

Venus flower basket (sponge), as
    Japanese wedding gift, 12
Vermeij, Geerat, 81, 83–84, *83*
vertebrates, 102, *102*, 103
    amphibians, 111
    body plan, 103
    dinosaurs, 78, 79, 111
    jaws, importance of, *106*, 108
    mammals, 111–113, *112*, *113*
    reptiles, 111
    See also chordates; fish

Walcott, Charles, 51, 53, 102
Walcott's Quarry, 52–53, *52*
    see also Burgess Shale
Wallace, Sally Mae, 35
Ward, Peter Douglas, 78, 79, 87, 89
water flea, *20*
Wedgwood, Josiah, 49
Wells, Martin, 38, 67
whelks, Kellett's, *80*
Whittington, Harry, 53
Wilson, E. O., 101
worms. *See* annelids; earthworms;
    flatworms; tapeworms

zebras, *112*
zooxanthellae, 27

PHOTOGRAPHY AND ILLUSTRATION CREDITS

All photos and animation stills from the video series "The Shape of Life" except:

The Boxwood Press: 20. ▫ Caira, Janine, Kirsten Jensen and Claire Healy: 43 right top, middle, bottom. ▫ Cambridge University Library, by permission of the Syndics: 4. ▫ Carlson, Kirsten: 10, 20, 23, 41, 78-79, 117. ▫ Caudle, Ann/ Monterey Bay Aquarium: 3, 21, 66, 67, 86, 88 bottom, 94. ▫ Dover Pictorial Archives Series: ix, 1, 12 bottom right, 19, 31 bottom right, 35, 38 left, 49, 59, 61, 63, 64, 71, 77, 78, 79, 85, 88, top left, 91, 93 top right, 101, 108, 109, 115, 119. ▫ Energy Film: 7 right, 75 bottom. ▫ Flowers, A. and L. Newman: 46, 47 bottom right. ▫ Hall, Howard/howardhall.com: 8, 9. ▫ Hickman, Dr. Carole: 82 top. ▫ Krautter, Manfred: 12 left top, right middle, bottom. ▫ Langstroth, Libby and Lovell: xii, 1, 8 right bottom, 9, 47 top left, 60 bottom left, middle, 105. ▫ Lanting, Frans: 112, 113 top. ▫ Mills, Claudia/Monterey Bay Aquarium: 25. ▫ Monterey Bay Aquarium Research Institute/Steve Haddock: 33 2nd row bottom right ▫ National Geographic Television: 43 left top, bottom, 75 top, 102, 106, 107, 108 top, 109 top left, right, bottom left, right, 113 bottom, 114, 118, 124. ▫ Pisera, Andrzej: 12 top right, bottom. ▫ Racicot, Craig/Monterey Bay Aquarium: 33 2nd row left. ▫ Reiswig, Henry: 12 left middle, bottom. ▫ Seaborn, Charles: 2 top, 8 left middle, bottom, 34, 35, 47 top right, bottom left, 60 top middle, middle, bottom left, 76, 77. ▫ Seaborn, Charles/Monterey Bay Aquarium: iv, xi, 8 top left, right, middle bottom, 18, 19, 22, 28, 60 top left, 62, 63, 65, 70, 80, 81, 92, 93, 98, 99, 100, 101. ▫ Sea Studios, Inc.: 48, 89, 90, 91. ▫ Shear, William: 72 right top, bottom. ▫ Smithsonian Institution Archives, Charles D. Walcott Collection: 51, 55. ▫ University of California, Santa Barbara, and American Chemical Society: 84. ▫ Vermeij, Dr. Geerat: 83 top right. ▫ Webster, Steven K., 2 bottom, 8 right middle, 60 top right. ▫ Wrobel, Dave/Visual Fun, Ltd.: ii-iii, vi-vii, 31. ▫ Wrobel, Dave/Monterey Bay Aquarium: 23 bottom, 33 top row, 2nd row right, 3rd and 4th rows.